Introduction to Synthetic Polymers

Ian M. Campbell

School of Chemistry,
University of Leeds

Oxford New York Tokyo
OXFORD UNIVERSITY PRESS
1994

University Press, Walton Street, Oxford OX2 6DP

Oxford   New York   Toronto
Delhi   Bombay   Calcutta   Madras   Karachi
Kuala Lumpur   Singapore   Hong Kong   Tokyo
Nairobi   Dar es Salaam   Cape Town
Melbourne   Auckland   Madrid

and associated companies in
Berlin   Ibadan

Oxford is a trade mark of Oxford University Press

Published in the United States
by Oxford University Press Inc., New York

A catalogue record for this book is available from the British Library

Library of Congress Cataloging in Publication Data
Campbell, Ian M. (Ian McIntyre), 1941–
Introduction to synthetic polymers / Ian M. Campbell.
(Oxford chemistry series ;)
Includes bibliographical references and index.
1. Polymers.   I. Title.   II. Series.
QD381.C34   1994   547.7–dc20   93–1953

ISBN 0-19-855399-4
ISBN 0-19-855398-6 (pbk.)

Typeset by Colset Private Limited, Singapore
Printed in Great Britain by
Redwood Books, Trowbridge, Wiltshire

Preface

Ten or so lectures on synthetic polymers are a usual component of undergraduate chemistry courses. The available textbooks have usually been written for more specialist interests. These books, apparently full of 'thousands' of polymeric chain structures, can discourage the students and make it an uphill struggle to instil confidence that a reasonable understanding of the main principles underlying the synthesis and properties of polymers can be achieved.

The aim of this book is to use a relatively small number of polymers as the background for an introduction to the most important aspects of modern polymer science. The amounts involved in commerce are the main basis of the particular polymers which are used as examples. A book of this length must be somewhat selective, focusing on the more important facets. For example, books on polymers are often overburdened with the range of methods for molecular weight determination: here it is recognized that just a few of these cover the large bulk of relevant activity in modern laboratories. There is little need beyond this for a reader seeking an introduction to the topic.

There are two main questions of fundamental importance here. How do the properties of a polymeric substance in bulk relate to the sizes and architectures of the molecules in it? How may the desired features be imposed on polymeric chains during synthesis? To illustrate the possible variations for just one polymeric substance, it may contain chains which are all around the same length or which cover a substantial range of lengths; the chains may be linear (unbranched) or branched or, indeed, of more complex forms such as 'comb-like' or 'star-like'; there may be chiral (asymmetric) centres in the chains and in sequence these may be all the same, randomly different or alternating. What effects do these subtle differences produce at the macroscopic level and how may the syntheses of the different types of polymer be brought about?

Chapter 1 defines terminology before giving historical and industrial perspectives: it then introduces general aspects of polymerization processes. The averages and distributions of molecular masses of polymers, their definitions and main methods of measurement, are the topics of Chapter 2. The microscopic structures of bulk polymers and their effects on the physical properties are assessed in Chapter 3, whilst Chapter 4 goes on to discuss how modern techniques may be applied to obtain detailed information on microstructural features. Chapters 5–7 describe the major mechanisms of polymerizations, step-growth, radical-addition, and ionic/coordination-addition types respectively, using important industrial processes for exemplification. Chapter 8 relates various main features of chemical and physical structures to the properties of common polymers. It also considers the problems associated with polymer disposal. In the course of Chapter 9, a number of polymers of interest on account of their specialized usages rather than large-scale productions are brought in, such as those which are able to retain strength at high temperatures or which combine elasticity with remouldability. Finally, in Chapter 10 we peer into the future to some extent, considering some of the exciting new polymeric materials with astonishing properties which have become available recently.

A particular feature of this book is the appearance at the end of each chapter of a set of exercises, some drawing on recently published research papers. These are designed to allow readers to test their level of understanding and, in some cases to advance it. At the end of the book are selected suggestions for further reading.

It is hoped that readers will find it easier to gain an understanding of the basic principles of polymer chemistry and physics against this background of a limited number of different polymers. Thereafter, these principles may be carried forward into some of the more advanced textbooks with enough confidence to cope with the much larger number of different polymers usually found therein.

Leeds   I.M.C.
April 1993

Contents

1

Introduction

'Polymer' as a scientific term was used first by the famous Swedish chemist, J. J. Berzelius, in 1827. But almost a century elapsed thereafter before a properly controlled synthesis of a polymer was achieved. We are familiar with a range of synthetic polymers nowadays, even if most of these have only made their debut in the last 60 years or so. New polymers continue to be developed and 'plastics' can be expected to turn up in even more commercial products in the future.

1.1 Definitions and nomenclature

A *polymer* is defined as a substance composed of very large molecules (or *macromolecules*). The molecular structure corresponds to a *chain* composed of many small molecules joined by chemical bonds. In synthetic polymers one structure, termed a *constitutional repeating unit (CRU)*, can be considered as producing the polymer molecule by its replication. Each CRU is joined to its neighbours by covalent bonds which are part of the *backbone*, defined as the minimum set of bonds and intervening atoms which extend continuously from end to end of the polymer molecule. The *monomer(s)* is (are) the small molecule(s) of one or more sorts which are incorporated into the polymer as it is synthesized. A *monomer residue* results from the incorporation of one monomer molecule into the chain. In many cases the CRU and monomer residue may be identical, as with a vinyl polymer where both correspond to $-CH_2-CH(X)-$. But the CRU is larger than any monomer residue in other cases, such as *copolymers* which incorporate more than one monomer. Polymers show special properties in comparison with *oligomers* which have molecules composed of up to a few tens of CRUs only, rather than the at least 50 in a polymer molecule.

Synthetic polymers are made by reacting monomers or their derivatives under controlled conditions. When a single monomer is concerned, the result is commonly a *homochain* polymer composed of just one monomer residue: polyethylene $(-CH_2-CH_2-)$ and poly(vinyl chloride) (often known by its acronym PVC) $(-CH_2-CH(Cl)-)$ are examples. Other common polymers require two monomers. Poly(ethylene terephthalate) (acronym PET) has the CRU represented as $-O-CH_2-CH_2-O-C$

$(=O)-C_6H_4-C(=O)-$ and may be regarded as being the *alternating copolymer* of ethylene $(CH_2=CH_2)$ and terephthalate $(O(O=)C-C_6H_4-C(=O)O)$, where C_6H_4 represents a doubly substituted (1,4- or *para-* in this case) aromatic ring. In principle, the monomers for the synthesis of PET are ethylene glycol $(HOCH_2-CH_2OH)$ and terephthalic acid $(HO(O=)C-C_6H_4-C(=O)OH)$, repeatedly condensed together (or esterified) with elimination of a water molecule for each linkage. Thus PET is often termed a *polyester*.

The common names of monomers are used in most of the science and technology and essentially all of the commerce of polymers. This demands that this book should follow custom and most practice in *not* using the systematic names of organic compounds. Corresponding names are listed for monomers in Appendix 1.

All polymers have a very long backbone and this is the only macromolecular feature of a *linear* or *unbranched* chain. Evidently *branched* polymers have other chains joined to the backbone at various points along its length. *Crosslinks* are short linkages, commonly involving just a few covalent bonds, between polymer chains. Figure 1.1 represents these types of chain features in a simple way in two-dimensional projection. In considering how the chain backbones, such as the carbon atoms of polyethylene (shown as irregular lines here) are represented, it should be realized that

Linear chain

Branched chain

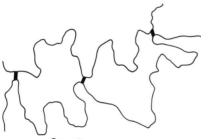

Crosslinked chains

Fig. 1.1. Simple representations of linear, branched, and crosslinked chains (crosslinks are indicated by short thick lines).

individual chains are tangled with one another in the typical bulk sample. An apt description might be 'polymer spaghetti'.

At the general level, there are several types of copolymer. Consider two monomer residues designated as A and B. Figure 1.2 shows microscopic sequences and identifying class names. Any monomer must have at least two functional groups which have reacted or at least one double bond which has become a single bond when the corresponding residue has been incorporated into the polymer, i.e. the monomer must be *bifunctional, trifunctional* or occasionally *tetrafunctional*. Many common monomers are bifunctional, such as vinyl chloride in producing PVC. Some monomers can be trifunctional, giving monomer residues which each have *up to* three covalent linkages to other residues, as is seen for A in the network copolymer in Fig. 1.2. *Network copolymers* evidently require that one of the corresponding monomers has a functionality greater than 2.

One of the major interests explored in this book will be how structural differences at the microscopic level result in different properties of bulk samples of polymers, i.e. at the macroscopic level. The response to mechanical stress is one basis of differentiation. Some bulk polymers are *rigid*; that is to say that the shape does not deform significantly under moderate applied stress. Other polymers are *flexible*, so that they deform noticeably (for example bend) but will recover slowly and sometimes only partially on release. Another class is composed of *rubbery* or *elastic* solids, distinguished by the abilities to be deformed quite severely (for example

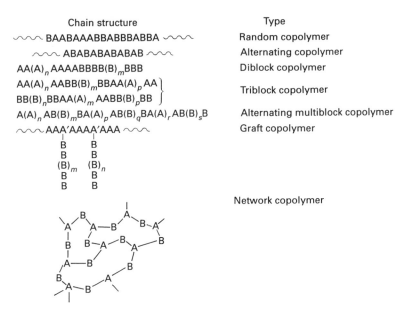

Fig. 1.2.   Types of AB copolymers, where A and B are different monomer residues.

stretched) by quite small applied stress and to 'snap back' quickly to regain their original shape on release. A lump of a 'rubber' or *elastomer* (i.e. a polymer showing elastic properties) gives to a prod from a finger (say) far more than would a lump of polyethylene, a typical flexible solid.

There are also obvious differences between some of the polymers which are rigid at ambient temperatures when they are heated. Network copolymers are more commonly referred to as *thermoset(ting) resins*; these do not soften at all before the temperature becomes high enough to induce thermal decomposition. Polymers of the largest class, *thermoplastics*, become flexible above a particular temperature, becoming rigid again on cooling through this temperature: this rigid ⇌ flexible cycle can be repeated on reheating and cooling again. In its flexible condition a thermoplastic is solid but can be moulded into complex shapes which are preserved in the rigid state after cooling. Remoulding can be done after reheating. Thermoplastics can be subdivided on the basis that only some are *fibre-formers*, which means that they can be *drawn* (i.e. pulled out) into strands which are permanent unless the polymer is melted and have considerable tensile strength and durability. PET is an example, giving rise to fibres known under various trade names such as Terylene (ICI) and Dacron (Du Pont).

1.2 Some historical perspectives

Natural polymeric materials, such as wood and horn, have been used by humans since prehistoric times. Modified forms of natural polymers were produced during the last century: the dramatic improvement in the properties of natural rubber (such as loss of tackiness and gain of elasticity) which resulted from its heating with sulphur, a process now referred to as *vulcanization*, was discovered independently in the USA (Goodyear in 1839) and the UK (Hancock in 1843). Modified forms of cellulose, initially cellulose nitrate (nitrocellulose) and later cellulose acetate and reconstituted cellulose itself were produced. But none of these can be considered as truly synthetic polymers.

The first fully synthetic polymers were the phenol–formaldehyde thermoset resins prepared by Baekeland in the early years of this century. These resins, known as Bakelites, achieved their main commercial significance in the 1920s. Several of today's most familiar polymers made their first appearances in Du Pont laboratories in the 1930s. Dashes of serendipity are apparent in the brief accounts of the discoveries which follow.

Carothers is the name associated with the first appearance of what we now know as nylon 6,6. At first, all that seemed to have come from the synthesis was a sticky lump, for which it did not seem worth seeking a patent. The true value only became evident when a glass rod was inserted and withdrawn, pulling silky fibres of remarkable tensile strength out from the lump. A patent was then sought with alacrity!

The reaction of hydrogen chloride with vinylacetylene ($CH_2=CH-C\equiv CH$) was being investigated and the critical event occurred reportedly when a sample of the resultant liquid was inadvertently left exposed to the air over a weekend. This had become solid by the time it was seen again and vitally important was the fact that the material was dropped and observed to bounce. Here then was the debut of the first synthetic rubber, a material now known generally as neoprene (see Exercise 1.3 at the end of this chapter).

Plunkett, with Du Pont in 1940, is credited with the discovery that tetrafluorethylene ($CF_2=CF_2$) can be polymerized; many failed attempts had led to the consensus that this monomer was unpolymerizable. The critical point came when a new cylinder of the monomer gas would not deliver any, even though the weight was correct for 'full'. Plunkett's curiosity persisted to the extent that he had the cylinder sawn open, when

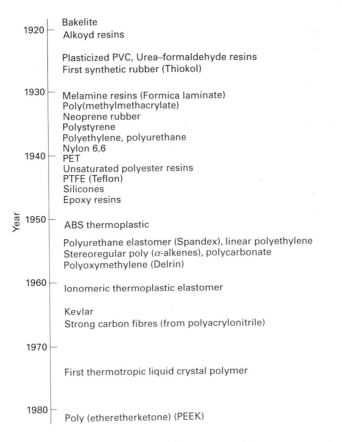

Fig. 1.3. Chronological list indicating the first *commercial* appearance of various polymers.

a white waxy powder was discovered within, thus exposing polytetrafluoro-ethylene (PTFE) to human gaze for the first time. This soon spurred the renewed effort which has now resulted in PTFE or Teflon being widely applied to produce nonstick finishes. In fact Plunkett and his coworkers ran an unrealized risk in their procedure of preparing tetrafluoroethylene and keeping it in steel cylinders. These cylinders were potential bombs since spontaneous conversion of the monomer to carbon and tetrafluoromethane liberates around one third of the energy released by the same molar amount of TNT in an explosion (see Exercise 1.6 at the end of this chapter).

Figure 1.3 completes this brief review of the history of synthetic polymers by indicating the chronological order in which various polymers made their first *commercial* appearance.

1.3 The synthetic polymers industry

Each year some 7×10^{10} kg of synthetic polymers are produced world-wide. Figure 1.4 shows the dominance of the industrial nations in this con-nection. The large production in the USA is evident. Table 1.1 indicates the contributions of the various types to total production of synthetic polymers by weight in the USA.

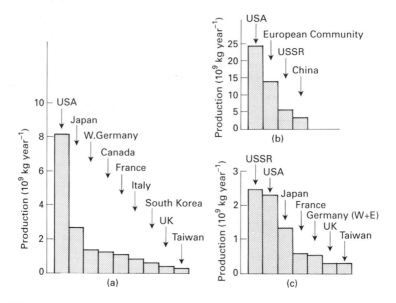

Fig. 1.4. Amounts of polymers made in the main producing nations in 1989. (a) Polyethylene (low and high density types), (b) all plastics and synthetic fibres, and (c) synthetic rubbers. Based upon entries given in 'Facts and figures for the chemical industry' given in *Chemical and Engineering News*, 18 June 1990, Volume **68**, Number 25, pp. 34–83.

Table 1.1 Main components of total polymer production in the USA

Type	Fraction of total (%)	Main specific polymers (Fraction of total)
Thermoplastics (non-fibres)	68	Polyethylene (23%) Poly(vinyl chloride) (13%) Polypropylene (11%) Polystyrene (8%)
Thermoplastics (fibres)	14	PET (5%) Nylon 6,6 (4%) Poly(alkenes) (2%)
Thermoset resins	10	Phenol–formaldehyde (5%) Urea–formaldehyde (2%)
Synthetic rubbers	8	SBR (2%) Poly(butadiene) (1%)

Based upon data given in *Chemical & Engineering News*, 18 June 1990, pp. 40-1.

It is evident that a small number of polymers dominate these production statistics: a similar situation can be expected in other industrial nations.

The capital cost of bringing a new thermoplastic into commercial production is enormous. Two to three thousand million pounds sterling (three to five thousand million US dollars) is a reasonable estimate for bringing on stream a new monomer production facility, new designs of polymer production plants and novel types of processing and fabrication operations. On this basis it can be expected that there will be relatively little change in the relative proportions of the amounts of main polymers produced on a national or global scale in the near future. This is supported by profiles representing the histories of annual productions of main polymers in the USA appearing in Figs. 1.5–1.7. Synthetic rubbers are not accorded similar representation since little change has occurred. Total productions in the years 1969 and 1989 were in fact not significantly different and there are only minor fluctuations during the intervening years.

It is obvious from Fig. 1.5 that production of thermoplastics in the USA has been increasing steadily, trebling over around 20 years in fact. Although the 'oil price shocks' of 1974–5 and 1979–80 make clear impacts on the profiles, apparently they have not disturbed the longer term trends. There have been almost parallel rises in the main individual thermoplastics except that two, high density polyethylene and polypropylene, have risen noticeably faster than the others. In Chapter 7 it will become apparent that this reflects the widespread commissioning of production plants within which relatively novel catalysts are used to synthesize these polymers.

Overall production of synthetic fibres has shown lower growth than total production of thermoplastics, several of the individual polymers being

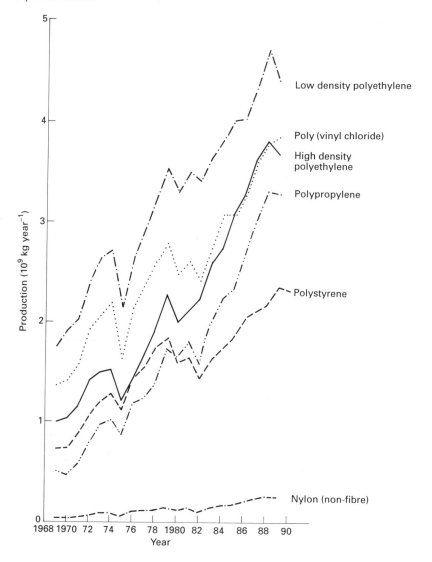

Fig. 1.5. Profiles of annual productions of the major thermoplastics in the USA from 1969 to 1989. Based upon entries given in the annual statistical surveys titled 'Facts and figures for the chemical industry' presented each year from 1970 in *Chemical and Engineering News*, terminating with the issue dated 18 June 1990, Volume **68**, Number 25, pp. 34–5.

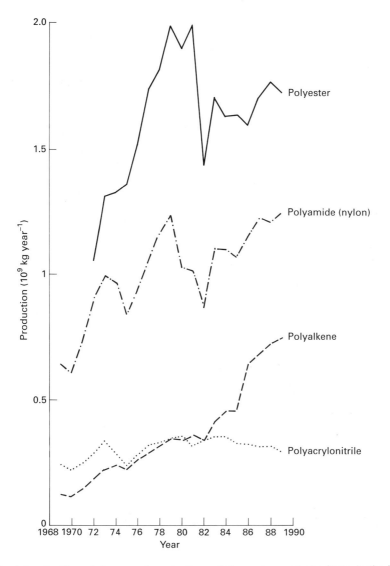

Fig. 1.6. Profiles of the annual productions of the major synthetic fibres in the USA from 1969 to 1989. Source as for Fig. 1.5.

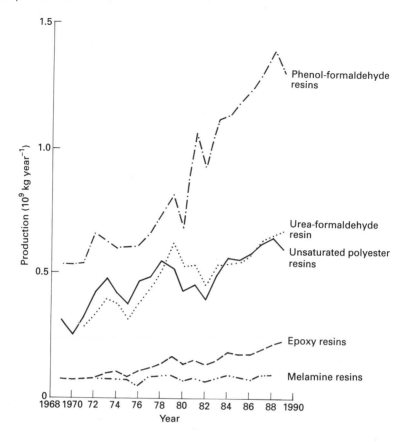

Fig. 1.7. Profiles of the annual productions of the major thermoset(ting) resins in the USA from 1969 to 1989. Source as for Fig. 1.5.

common to both classes. Polyalkene (principally polypropylene) fibre production has shown the most spectacular growth, around a factor of six in two decades, but started from a low level. Polyester (PET principally) and polyamide (or nylon) (dominated by nylon 6,6) compose the main synthetic fibres.

Figure 1.7 shows that total production of thermoset resins has roughly doubled over two decades. Evidently phenol–formaldehyde is the predominant individual resin, a position enhanced by recent growth rates.

The synthetic polymers industry is a major consumer of the output of the chemical industry. Each year a list of 'the top 50 chemicals in the USA' is published, in which several of the entries reflect usages as monomers. Table 1.2 gives relevant entries with rankings and indicates the main polymeric products.

Table 1.2 Relevant entries in the 'top 50 chemicals' list

Rank	Chemical	Main polymer product(s)
4	Ethylene	Polyethylene
9	Propylene	Polypropylene
18	Vinyl chloride	Poly(vinyl chloride)
20	Terephthalic acid	Poly(ethylene terephthalate)
21	Styrene	Polystyrene
23	Formaldehyde	Urea/phenol thermoset resins
26	Ethylene glycol	Poly(ethylene terephthalate)
28	Ethylene oxide	Poly(ethylene oxide)
33	Phenol	Phenol-formaldehyde resins
37	Butadiene	Synthetic rubbers
39	Acrylonitrile	Polyacrylonitrile, polyacrylamide
41	Vinyl acetate	Poly(vinyl acetate), poly(vinyl alcohol)
47	Adipic acid	Nylon 6,6
50	Caprolactam	Nylon 6

Based upon data given in *Chemical & Engineering News*, 18 June 1990, p. 37.

Several other entries in the list owe their rankings substantially to usage as precursors for monomers. Examples are benzene (ranked 16), ethyl benzene (19), and cyclohexane (42).

The polymers mentioned in this section will be used for exemplification of principles and procedures as much as possible in this book.

1.4 General features of polymerization processes

Two general types of chemical processes are responsible for the incorporation of monomers into a polymeric chain, referred to as *step-growth* and *addition* mechanisms.

In step-growth polymerization, the elementary acts are reactions between molecules and by repetition long chains are gradually produced. In the commonest cases, the elementary chemical reaction is a *condensation* which results in the elimination of a small molecule in the course of creating the new chemical bond. This is the action in syntheses of polyesters and polyamides. Figure 1.8 represents the overall achievements of four step-growth polymerizations of major commercial significance. The latter pair of processes in this figure though make the point that step-growth polymerizations may proceed by repeated reactions which are not condensations because no molecules are eliminated.

The number coding of the nylons indicates their molecular structure: the number(s) indicate how many directly-linked carbon atoms there are in the CRU (represented within the brackets in Fig. 1.8). In the CRU of nylon 6,6 there are two sequences of six carbon atoms each separated by a nitrogen

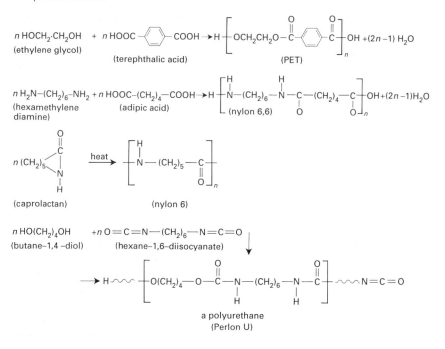

Fig. 1.8. Overall equations for the syntheses of major step-growth polymers, showing the monomers concerned in principle.

atom. The first number refers to the amine residue: thus nylon 6,10 has a CRU composed of the residues of $H_2N(CH_2)_6NH_2$ and $HO(O=)C(CH_2)_8C(=O)OH$.

In contrast, the elementary acts in addition polymerization are reactions between *monomer* molecules and the reactive end of a growing polymer chain. The addition of the monomer molecule simply results in the transference of the reactive site to the new chain end, which is then ready to add on another monomer molecule. It is thus apparent that addition polymerization is in fact a *straight (or linear) chain reaction*; the *chain carrier* is viewed collectively as all actively growing entities. The *propagation step*, in the conventional terminology of chain reactions, results in the relatively rapid addition of one monomer molecule. Monomer molecules will be abundant in typical addition polymerization systems: hence once a reactive site (such as a free radical centre) has appeared in the system (through *initiation* in chain reaction terminology), it can induce a large number of propagation acts in a very short time usually. Thus very long polymer chains can be generated in the earliest stages of addition polymerizations, in contrast to step-growth systems where molecules of any length react together rather slowly so that long chains can only appear late in the process.

The monomers which are most commonly polymerized by addition mechanisms are so-called vinyl compounds, of general formula $CH_2=CH(X)$. The overall result of polymerization may be represented simply as

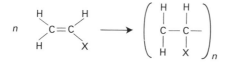

Vinyl chloride (X=Cl), vinyl acetate (X=$CH_3C(=O)O$), and acrylonitrile (X=CN) are important monomers. The monomer residues are shown above to be joined in a *head-to-tail* configuration, when the substituents X are attached to every second carbon atom in the backbone of the polymer molecule. The *configuration* of a carbon atom in a polymer chain cannot be changed without breaking chemical bonds and forming new ones and is thus fixed in the synthesis. The alternative head-to-head configuration results in relatively bulky X atoms or groups being attached to neighbouring carbon atoms and is thus much less likely in general because of greater steric interference and/or dipolar repulsion effects. The head-to-tail configuration can be taken as corresponding to the overwhelming proportion of linkages between monomer residues in vinyl polymers.

There is a second important feature in vinyl polymers of this type; every carbon atom in the backbone attached to X is a *chiral centre* through having four different substituents, two of which are the other parts of the polymer chain. Thus each monomer residue may have one of two different *stereoisomeric configurations* (conventionally designated as 'd' and 'l') determined by the relative disposition in space of the H and the X bonded to the chiral carbon atom. The *tacticity* of the polymeric chain is defined by the relative stereoconfigurations of chiral centres along the chain. When all are the same (say all d) the polymer is said to be *isotactic*. When every chiral centre has the opposite configuration to those of its neighbours, corresponding to strictly alternating d and l centres along the chain, the polymer is *syndiotactic*. When the sequence is random, the polymer is *atactic*. Figure 1.9 represents parts of the chains of the three types; for simplicity the chains are shown in the linear zig-zag conformations which puts all of the backbone carbon atoms into one plane. It is rather rare however for a polymeric chain to be completely stereoregular (for example 100 per cent isotactic) so that another system is required which specifies the degree of stereoregularity of chains. Consider a successive pair of CRUs of the type $-CH_2-CH(X)-$, which are said to constitute a *dyad*. This is

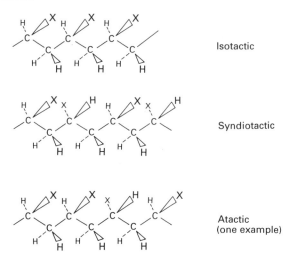

Isotactic

Syndiotactic

Atactic
(one example)

Fig. 1.9. Representations of the planar zig-zag conformations of stereoisomeric chains composed of the monomer residue $-CH_2-CH(X)-$.

labelled as 'm' when both chiral atoms have the same configuration and as 'r' otherwise. *Triads* of three successive CRUs can then be considered, which can be 'mm', 'mr' or 'rm', or 'rr'. Evidently a particular proportion of isotactic nature implies certain percentages of the dyads being m and the triads being mm. A fully atactic nature must imply equal numbers of m and r dyads and hence triads in the proportions of $1:2:1$ for mm, mr + rm, and rr for another instance. The system develops to *tetrads* and *pentads*. The sequences in triads, tetrads and some pentads are represented in Fig. 1.10 using X above C to represent one of 'd' or 'l' and X below C for the other.

Now consider the act of propagation and in particular what has to happen in the conversion of a monomer molecule into a monomer residue in a vinyl polymer. Looking at the basic process for ethylene addition to a growing chain

$$H_2C=CH_2 \rightarrow -CH_2-CH_2$$

it is necessary to regard the π part of the double bond as breaking first, which requires an enthalpy input of $260\,kJ\,mol^{-1}$. Then a new C−C σ-bond must form, attaching the new monomer residue and releasing enthalpy equivalent to $365\,kJ\,mol^{-1}$. These enthalpy terms combine to yield a net exothermicity ($\equiv -\Delta H$) of $105\,kJ\,mol^{-1}$, which indicates the main driving force for polymerization. The entropy change (ΔS) for polymerization is evidently negative (number of separate molecules decreases) so that in the equation $\Delta G = \Delta H - T\Delta S$ the last term is positive and thus makes a contribution to ΔG which corresponds to resistance

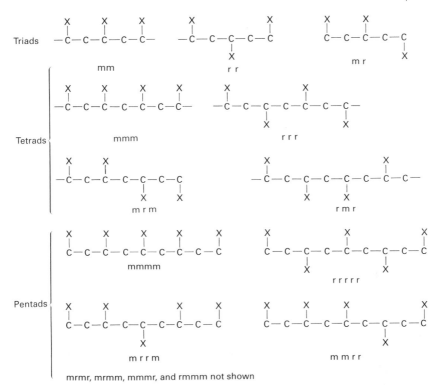

Fig. 1.10. Representations of triads, tetrads, and some pentads of various configurations, indicating the descriptions in mr notation, for the polymer chains composed of $-CH_2-CH(X)-$ residues. The hydrogen atoms are not shown and the two stereoisomeric configurations at the chiral carbon atom are represented by X above and below C respectively.

to polymerization. It follows that there will be some *ceiling temperature* (T_c), above which ΔG is positive and thus depolymerization is the favoured process. With the majority of vinyl polymers T_c is high enough to be of little concern for the synthesis of the polymer. But there are a few polymers which have low enough T_c to result in the polymer only being synthesized by a route which is associated with reasonably rapid polymerization at relatively low temperatures. Figure 1.11 indicates ceiling temperatures for a range of monomers when the pure liquids are polymerized, under pressurized conditions as required.

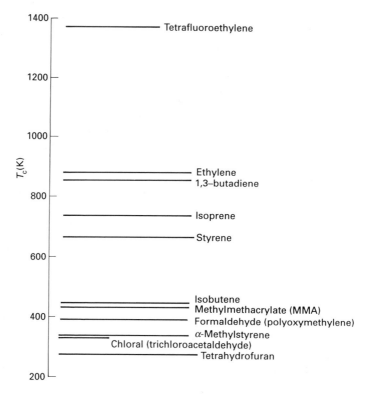

Fig. 1.11. A temperature scale indicating the values of the ceiling temperatures of polymers of the monomers listed.

Exercises

1.1. Match each of items (i)–(vii) in the left-hand column with the most closely associated one from the right-hand column

(i) alternating copolymer	(a) chain of 25 CRUs
(ii) elastomer	(b) polytetrafluoroethylene
(iii) oligomer	(c) $-[-CH(CH_3)-CH_2-]_n-$
(iv) Teflon	(d) nylon 6,6
(v) polypropylene	(e) dldldl $----$
(vi) polyacrylonitrile	(f) vulcanized rubber
(vii) syndiotactic polymer	(g) $CH_2=CH(CN)$ monomer.

1.2. A typical polystyrene used for moulding/extrusion has chains of average relative molecular mass of 1.2×10^5. How many monomer residues ($-CH(C_6H_5)-CH_2-$) would be required to give the same relative molecular mass in total? (relative atomic masses are 12.011 (C) and 1.008 (H)).

1.3. Write down overall equations representing the processes for the syntheses of the monomer from the addition of hydrogen chloride to vinylacetylene and the resultant polymerization to give neoprene which has the CRU $-[-CH_2-CH=CCl-CH_2]-$.

1.4. Using the system of Fig. 1.10, draw the remaining pentads. Predict the percentage abundances of isotactic triads, tetrads, and pentads in a homochain polymer having random distributions of m and r dyads.

1.5. Evaluate the enthalpy change (ΔH) and entropy change (ΔS) for the addition polymerization of one mole of methylmethacrylate and predict the ceiling temperature for poly(methylmethacrylate) when the thermodynamic parameter changes in the propagation step correspond to $\Delta H = -0.556 \text{ MJ kg}^{-1}$ and $\Delta S = -1.28 \times 10^{-3} \text{ MJ kg}^{-1} \text{ K}^{-1}$.
(Methylmethacrylate is $CH_2=C(CH_3)C(=O)OCH_3$ and the relative atomic masses are 1.008 (H), 12.011(C), and 16.000 (O).)

1.6. Enthalpies of formation at 298 K (ΔH_f) are as listed below. Use these to justify the statement that the thermal decomposition of tetrafluoroethylene (C_2F_4) to graphite and tetrafluoromethane (CF_4) liberates about a third of the energy evolved in the combustion of the same amount of trinitrotoluene (TNT) ($CH_3C_6H_2(NO_2)_3$) to yield CO_2, H_2O and N_2.

Substance	$CF_4(g)$	$CO_2(g)$	$H_2O(g)$	$C_2F_4(g)$	TNT(s)
$-\Delta H_f(\text{kJ mol}^{-1})$	936	394	242	659	2583.

1.7. The enthalpy change in the polymerization of methylmethacrylate has been measured as $-57.0 \text{ kJ mol}^{-1}$. The ceiling temperature for the polymerization of methylmethacrylate is 434 K. What is the corresponding entropy change when one mole of methylmethacrylate polymerizes?

2
Average molecular masses and polydispersity

2.1 Definitions and illustrations

Polymers are generally *polydisperse*, meaning that in a sample the individual molecules are not all of the same size and there is a range of molecular masses accordingly. Thus only *average* values of relative molecular masses can be specified usually for a bulk polymer. Immediately there is complexity because there are several types of average: which of these averages is measured depends on which method is used to determine it.

The simplest average is the *number average relative molecular mass* denoted as $\bar{M}_n$. Consider a particular sample of a polymer and imagine that by some means it could be completely fractionated, so that each separate fraction consisted of a specific number, say N_i, of molecules of just one relative molecular mass, say M_i. No two fractions have the same values of M_i and every molecule in the original sample is included in the relevant range of M_i values. The total number of molecules in the original sample is then specified by ΣN_i, which by implication covers all possible values of the variable i. The relative mass of N_i molecules of relative molecular mass M_i is the product $N_i M_i$, so that $\Sigma N_i M_i$ is the mass of the original sample. $\bar{M}_n$ is the total mass divided by the total number of molecules, expressed as

$$\bar{M}_n = \sum N_i M_i \Big/ \sum N_i \qquad (2.1)$$

Consider a sample consisting of just four fractions — although this is evidently unrealistic for polymers, this example is easily appreciated and reveals fundamental points. Table 2.1 sets out the values. The summations of N_i and $N_i M_i$ on the right combine according to eqn (2.1) to yield $\bar{M}_n = 24.0 \times 10^8/6000 = 4.00 \times 10^5$.

The weight of the fraction with relative molecular mass M_i is specified as $w_i = N_i M_i$. The *weight average molecular mass*, denoted by $\bar{M}_w$, is defined by analogy with eqn (2.1) with w_i replacing N_i as the weighting factor in the averaging process. Thus it is expressed as

Table 2.1 Illustrative data for average molecular masses

N_i	1000	2000	2000	1000	$\Sigma N_i = 6000$
$10^{-5}M_i$	1.00	3.00	5.00	7.00	
$10^{-8}N_iM_i$	1.00	6.00	10.0	7.00	$\Sigma N_iM_i = 24.0 \times 10^8$
$10^{-13}N_iM_i^2$	1.00	18.0	50.0	49.0	$\Sigma N_iM_i^2 = 118.0 \times 10^{13}$
$10^{-12}N_iM_i^{1.75}$	0.56	7.69	18.8	16.9	$\Sigma N_iM_i^{1.75} = 44.0 \times 10^{12}$

$$\bar{M}_w = \sum w_iM_i \Big/ \sum w_i = \sum N_iM_i^2 \Big/ \sum N_iM_i. \tag{2.2}$$

The value of $\bar{M}_w$ is expected to be influenced more by heavier molecules and hence to be larger than $\bar{M}_n$. The summations of $N_iM_i^2$ and N_iM_i given

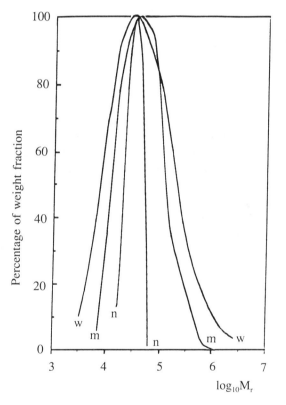

Fig. 2.1. Percentage distribution of weight fraction as a function of log M_r, where M_r is relative molecular mass, for polyethylene samples specified as 'narrow' (n) ($\bar{M}_w$ = 3.20×10^4, $\bar{M}_w/\bar{M}_n$ = 1.15), 'medium' (m) ($\bar{M}_w$ = 5.42×10^4, $\bar{M}_w/\bar{M}_n$ = 2.96), and 'wide' (w) ($\bar{M}_w$ = 1.55×10^5, $\bar{M}_w/\bar{M}_n$ = 10). Reprinted with permission from Barbalata, A., Bohossian, T., Prochazka, K., and Delmas, G. (1988). Characterization of the molecular weight distribution of high density polyethylene using the turbidity at a lower critical solution temperature. *Macromolecules*, **21**, 3286–95. Copyright (1988) American Chemical Society.

at the right of Table 2.1 combine to give $\bar{M}_w = 118.0 \times 10^{13}/24.0 \times 10^8 = 4.92 \times 10^5$, larger than the corresponding $\bar{M}_n$ value above. The fraction with the largest M_i contributes 42 per cent of $\Sigma N_i M_i^2$, 29 per cent of $\Sigma N_i M_i$, and only 17 per cent of ΣN_i, so that $\bar{M}_w$ must be larger than $\bar{M}_n$.

In a *monodisperse* sample, all molecules have the same value of M_i and thus $\bar{M}_w$ and $\bar{M}_n$ are equal, i.e. $\bar{M}_w/\bar{M}_n = 1$. In polymer systems $\bar{M}_w/\bar{M}_n$ is termed the *polydispersity index*; its value increases as the distribution of relative masses of molecules in the sample becomes wider, i.e. extends over a larger range of values of M_i. This is illustrated in Fig. 2.1 which shows profiles of weights of fractions (w_i) relative to the maximum w_i vs $\log_{10} M_i$ for samples of polyethylene described as aving 'small', 'medium', and 'large' polydispersities.

Another average is much used on the basis of its relative ease of measurement, $\bar{M}_v$ is the *viscosity average relative molecular mass* and is defined as

$$\bar{M}_v = \left\{ \sum N_i M_i^{(1 + \alpha)} \bigg/ \sum N_i M_i \right\}^{1/\alpha}. \qquad (2.3)$$

The parameter α is effectively a constant for a particular polymer dissolved in a specified solvent at a defined temperature and often has a value around 0.7–0.8 (see Table 2.2). When $\alpha = 1$, $\bar{M}_v$ becomes identical with $\bar{M}_w$. Otherwise $\bar{M}_v$ lies between $\bar{M}_n$ and $\bar{M}_w$, as is illustrated in the bottom row of Table 2.1, where $\alpha = 0.75$ is used: the summation of all $N_i M_i^{1.75}$ entries is 4.40×10^{13} and this leads to $\bar{M}_v = \{4.40 \times 10^{13}/24.0 \times 10^8\}^{1.33} = 4.68 \times 10^5$. This is between $\bar{M}_n = 4.00 \times 10^5$ and $\bar{M}_w = 4.92 \times 10^5$ as expected.

Table 2.2 shows illustrative values of these averages for some commercial samples of common polymers, together with the polydispersity indexes.

Table 2.2 Values of average relative molecular masses for some samples of commercial polymers

Polymer	$\bar{M}_n$	$\bar{M}_v$	$\bar{M}_w$	$\bar{M}_w/\bar{M}_n$
Polystyrene	1.33×10^5	2.14×10^5	3.36×10^5	2.52
	8.50×10^4	1.64×10^5	2.42×10^5	2.85
High-density polyethylene	1.50×10^4	4.60×10^4	1.47×10^5	9.80
	1.70×10^4	7.90×10^4	2.98×10^5	17.5
	1.60×10^4	7.60×10^4	3.91×10^5	24.4
Linear low-density polyethylene	1.20×10^4	4.20×10^4	4.80×10^4	4.00
	2.00×10^4	7.80×10^4	1.45×10^5	7.25
	1.00×10^4	7.60×10^4	1.31×10^5	13.1

2.2 Experimental determinations

Most textbooks on polymers give accounts of up to ten or so methods for the determination of average molecular masses. This reflects the historical background of polymer science rather than current practice largely. Nowadays research reports indicate two methods predominating; one of these, viscometry, may be viewed as available to 'the poor' in contrast with the other, gel permeation chromatography (GPC) (or size exclusion chromatography) which needs expensive equipment. A reasonable knowledge of these, along with a brief outline of one other, osmometry, is sufficient at the introductory level of this book.

2.2.1 Viscometry

This is the simplest technique and it can be performed with relatively unsophisticated and inexpensive equipment. It is versatile and can cope with highly corrosive solutions, for example concentrated sulphuric acid.

The method requires the measurement of the time (t) for a specified volume of a solution of polymer to flow through a length of capillary tubing: t is proportional to the coefficient of viscosity (η) of the solution. The solvent molecules are much smaller than the molecules of the polymer; these tangle together so that η increases measurably as the solutions become more concentrated in the range in which the density can be considered as unchanged from that of the pure solvent without introducing appreciable error. Ubbelohde viscometers are used commonly and Fig. 2.2 shows some designs which allow different extents of dilution. The typical dimensions of the capillary are a bore diameter of 0.2 mm and a length of 100 mm. The arms labelled 'A' contain the main elements, the capillary tubing, and the volume immediately above it provided largely by the lower of the two bulbs and defined by the two lines marked above and below this. The meniscus passes between these levels in time t as the solution flows through the capillary under the influence of gravity. At the start, a known volume of the most concentrated solution to be used is added through the arm C to lie in the main reservoir (labelled I, II or III). Air pressure is then applied down arm C and, with arm B closed off, the solution is forced up arm A until it has reached the upper of the two bulbs. Then the air pressure is released and arm B is opened, allowing that part of the solution below the capillary section of arm A to drain rapidly back into the main reservoir. This is the feature of the design which ensures that the total volume of liquid in the viscometer has no effect on the flowrate through the capillary. Thus successive measurements for increasingly dilute solutions of the polymer can be conducted without having to withdraw any liquid from the viscometer. With the injection of a known volume of additional solvent down arm C to the solution in the reservoir, the next solution for

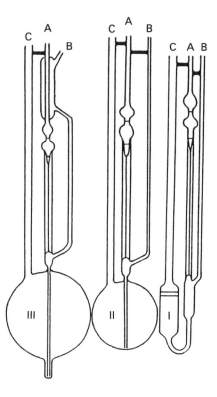

Fig. 2.2. Designs of Ubbelohde viscometers, with dilution of 24 cm^3 (I), 100^3 (II), and 250 cm^3 with extended base (III), when the minimum sample volumes are 6, 7, and 5 cm^3 respectively. Reprinted with permission from Mann, P. J., Wen, S., Xiaonan, Y., and Stevenson, W. T. K. (1990). A modified Ubbelohde viscometer with improved dilution characteristics. *European Polymer Journal*, **26**, 489–91. Copyright (1990) Pergamon Press.

measurement of the flow time (t) is ready after proper mixing. One other measurement is needed, the flow time (t_0) when the liquid is the pure solvent of viscosity coefficient η_0. For accurate work the viscometer should be within a thermostatically controlled enclosure.

A dimensionless parameter termed the *viscosity ratio* (η_r) is defined by

$$\eta_r = \eta/\eta_0 = t/t_0.$$

The *specific viscosity* (η_{sp}) is defined as equal to ($\eta_r - 1$) and gives rise to the *viscosity number*, defined as η_{sp}/c, where c is the corresponding mass concentration of the polymer in solution. Usually the values of the viscosity number increase as c increases, reflecting increasing entanglement of polymer chains with one another. It is the value extrapolated to infinite dilution, known as the *limiting viscosity number* or the *intrinsic viscosity*

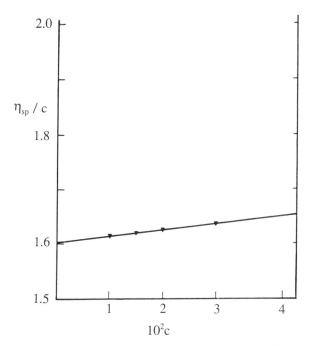

Fig. 2.3. Plot of specific viscosity (η_{sp}) divided by concentration, $c(\mathrm{kg\ m^{-3}})$ versus concentration of an ultrahigh molecular weight polyethylene dissolved in decalin at 408 K. Reproduced br permission of John Wiley & Sons Inc. from McHugh, A. J., Vrahopoulou, E. P., and Edwards, B. J. (1987). Molecular weight fractionation in tubular flow-induced crystallization of polyethylene, *Journal of Polymer Science: Part B*, **25**, 953–6. Copyright (1987) John Wiley & Sons Inc.

and defined as $[\eta] = \lim. (\eta_{sp}/c)_{c \to 0}$, which is of main interest. Figure 2.3 shows a representative plot of η_{sp}/c versus c, evidently fairly easy to extrapolate due to its apparently linear form.

The key to the use of viscometry here is that $[\eta]$ has an empirical relationship to $\bar{M}_v$ as expressed by the Mark–Houwink–Sakurada equation

$$[\eta] = k(\bar{M}_v)^\alpha. \tag{2.4}$$

The parameter α appeared in eqn (2.3). For a particular polymer/solvent/temperature combination, k and α are effectively constants, independent of $\bar{M}_v$ provided that an usually low value is not concerned, and are available in tabulations. A very extensive set of values appear in *The Polymer Handbook* (Brandrup, J. and Immergut, E. H. (ed.) (1989) 3rd edn, pp. VII, 5–46. Wiley, New York). Access to such tabulations can be considered as the means of making this technique an absolute method effectively for the determination of $\bar{M}_v$. Table 2.3 shows a selection of values of k and α over a range of systems at ambient temperature.

Table 2.3 Values of parameters for the Mark–Houwink–Sakurada equation for various polymer/solvent systems at 298 K

Polymer	Solvent	$10^6 k \, (\mathrm{dm^3 g^{-1}})$	α
Polystyrene	Toluene	8.6	0.74
	Chloroform	7.2	0.76
	Cyclohexene	16	0.68
Bisphenol-A-polycarbonate	Chloroform	30	0.74
	Dichloromethane	30	0.74
Poly(methylmethacrylate)	Benzene	5.5	0.76
Polyacrylonitrile	N, N-dimethylformamide	39	0.75
Nylon 6,3	Concentrated sulphuric acid	42	0.78
Nylon 6,6	m-cresol	240	0.61
Poly(vinyl acetate)	Butanone	42	0.62

The values of α in this table suggest that its 'normal' value is around 0.7. This is reasonable provided that two important 'deviations from normality' are borne in mind. The first of these is that 'low' values of α are expected when the solvent is relatively poor. For example polystyrene solutions in 2-heptanone, ethyl acetate, and decalin at 298K give values of $\alpha = 0.60$, 0.57, and 0.54 respectively, evidently significantly lower than the $\alpha \approx 0.75$ for good solvents of polystyrene, such as toluene and chloroform. On the other hand 'high' values of $\alpha (\sim 1.0)$ are found when the polymer has a rigid rod-like nature in solution. An example is poly(diisopropyl fumarate) which has a chain of $-CH(X)-$ backbone units with the pendant X being $-C(=O)-O-CH(CH_3)_2$, the sheer bulk of which appears to make the chain rigid in benzene solution at 300 K, as reflected by $\alpha = 0.98$. Also aromatic polyamides (Section 9.2) are characteristically extremely rigid, so that their solutions give rise to α values which often slightly exceed 1.0.

So, in summary, the 'poor person's way' to determine average relative molecular mass is viscometry, in conjunction with the extensive tabulations of values of the parameters for eqn (2.4).

2.2.2 Gel permeation chromatography (GPC)

This technique involves some expensive equipment of necessity and indeed modern instruments have integral computer facilities with dedicated programs routinely. But if this 'rich person's facility' can be afforded, it is undoubtedly the way to gain maximum information. A single run can yield all of the averages together with the distribution of relative molecular masses.

The central feature is a set of columns packed with insoluble gels, often composed of polystyrene crosslinked with divinylbenzene (see Fig. 8.4). These polymeric fillings of individual columns are semirigid beads synthesized so that they offer a uniform pore size when swollen with solvent. Commonly there will be four columns connected in series offering in turn nominal pore (or exclusion) diameters of 10^4, 1000, 100, and 50 nm typically. In operation, solvent is forced continuously through the sets of columns at a controlled rate (typically $10^{-3}\,dm^3\,min^{-1}$) by a high-pressure liquid pump. The sample of the polymer under investigation is injected in solution in a relatively small volume of this solvent at a position upstream of the columns. The detector is located downstream of the last column and responds sensitively to the presence of the polymer in a small volume of the eluent.

The key phenomenon is that individual polymer molecules explore the pore systems of the columns to the extent which is allowed by their size and hence relative molecular mass. Larger molecules are excluded from the smaller pores and can thus only diffuse into a restricted part of the pore system within the beads. In contrast the smallest molecules may have access to virtually all pores and thus spend a significantly longer time out of the main solvent flow between the beads than do the larger polymer molecules. Thus the largest molecules have the least residence time within the set of columns as a whole and emerge first, followed by progressively smaller molecules as elution time increases. Figure 2.4(a) represents at a simple level the origins of the different residence times of large and small molecules in a main channel with narrow pores to the sides. Figure 2.4(b) represents the course of separation of a hypothetical sample composed of 12 molecules of three sizes.

A variety of detection techniques are used; some of those most commonly used depend upon differences in refractive index, ultraviolet or infrared absorption or light scattering when polymer molecules are passing the detector station. Nuclear magnetic resonance (NMR) spectroscopy has also been used with great advantage, as will be discussed shortly. The primary result is a profile corresponding to the variations of some property which is characteristic of the polymer concerned versus elution time. It is then apparent that the GPC method is dependent upon calibration, ideally using samples of the same polymer having known and narrow distributions of molecular mass. In fact only a few polymers are available commercially as sets of such 'standardized' samples.

Polystyrene is the commonest of these; sets of samples can be purchased with individual relative molecular masses covering the range 600 to 3×10^6, which is of main interest. Polyethylene and poly(ethylene oxide) can also be obtained as sets of standardized samples for calibrating GPC systems. But it is evident that a procedure must be available which allows one polymer to be used to calibrate the system for another in general.

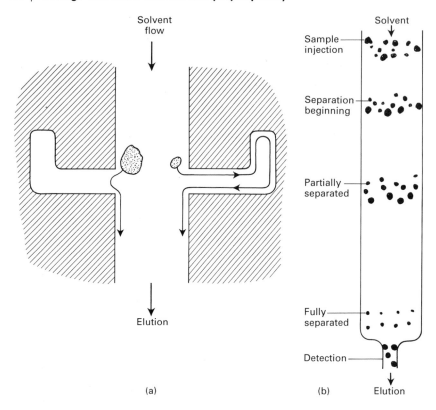

Fig. 2.4. Simple illustrations of the principle of gel permeation chromatography (GPC).

Viscometry in fact provides the basis of what is known as the *universal calibration method*: it has been shown empirically that the product of the intrinsic viscosity $[\eta]$ and the relative molecular mass, M, of a polymeric fraction depends only on the total volume (V) of solvent which has been eluted before this dissolved fraction emerges from the GPC column. Most importantly $[\eta]M$ is independent of the identity of the polymer, so that a *universal calibration plot* can be constructed using standardized samples of say polystyrene. Usually it is log ($[\eta]M$) which is plotted against V or elution (retention) time (t_R); t_R is directly proportional to V for a constant flowrate of solvent. Figure 2.5 illustrates this. The lower part shows a plot of $\log_{10}[\eta]$ versus $\log_{10} M$ for a series of polystyrene samples with very narrow molecular mass distributions (i.e. effectively monodisperse so that every molecule can be considered to have the relative molecular mass M): the linearity is expected from eqn (2.4). The upper part is the corresponding universal calibration curve. A polydisperse sample of another polymer will give an extended profile of detector response versus V (or t_R) as the

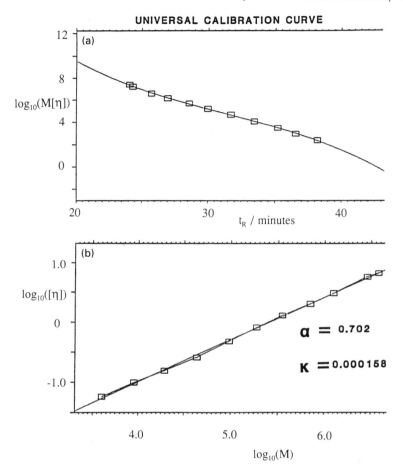

Fig. 2.5. (a) Universal calibration curve on a plot of $\log_{10}(M[\eta])$ versus retention time (t_R), where the locations of the points (empty squares) were determined using polystyrene standards of narrow weight distribution centred on relative molecular mass M. (b) Log–log plot of intrinsic viscosity ($[\eta]$) versus M for polystyrene standards in tetrahydrofuran at 313 K: the resulting values of parameters in the Mark–Houwink–Sakurada equation (eqn (2.4)) are shown within the diagram. Reprinted from Kou, C-Y., Provder, T., and Koehler, M. E., (1990). Evaluation and application of a commercial single capillary viscometer system for the charaterization of molecular weight distribution and polymer chain branching. *Journal of Liquid Chromatography*, **13**, 3177–99 by courtesy of Marcel Dekker Inc.

primary result of GPC analysis. The need is to convert this into the corresponding profile of number of molecules (N_i) versus relative molecular mass (M_i). The main problem is the conversion of the scale of V (or t_R) into the M_i scale. Suppose that a particular value of $V = V_i$ is selected for which the corresponding component of the polymer has relative molecular

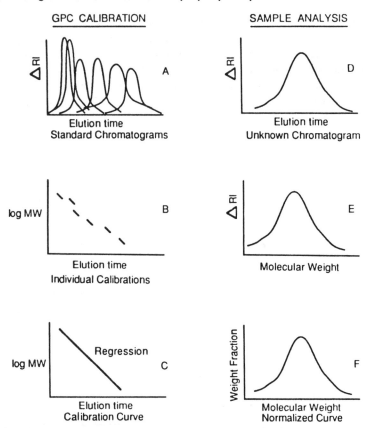

Fig. 2.6. Representation of the stages in the analysis of a sample of a polymer for its weight fraction distribution by gel permeation chromatography (GPC) using a refractive index (RI) detector. A–C represents the establishment of the universal calibration curve of the logarithm of the relative molecular weight (log MW) versus elution time. D represents the detected profile of the difference in RI (ΔRI) compared to the pure solvent versus elution time, from which E and F are derived using calibrations. Reprinted by permission of John Wiley & Sons Inc from Lee, C. H. and Mallinson, R. G. (1989). Molecular weight distribution in continuous stirred tank vinyl acetate emulsion polymerization. *Journal of Applied Polymer Science*, **37**, 3315–27. Copyright (1989) John Wiley & Sons Inc.

mass $M_i(\equiv M)$ and intrinsic viscosity $[\eta]_i$. Let K_i be the value of the product $[\eta]_i M_i$; its value can be read off the universal calibration plot at $V = V_i$. Moreover $[\eta]_i$ can be expected to be related to M_i by an equation of the form of eqn (2.4), i.e. $[\eta]_i = k(M_i)^\alpha$ and thus $K_i = k(M_i)^{(1+\alpha)}$. Since values of k and α are usually obtainable from tabulations, this allows evaluation of the value of M_i which corresponds to K_i and hence to V_i. This is the outline of a conversion process which is usually effected via the computer interfacing of the detection system. Figure 2.6 illustrates in simplified terms the procedures of calibration and extraction of a profile of weight fraction (w_i) versus M_i when detection is via the change in the refractive index (ΔRI) which is in fact proportional to the mass concentration of the polymer in the detected volume. It is apparent in the upper part of Fig. 2.5 that the central part of the universal calibration plot is linear to a good approximation. Hence the following set of equalities is expected (where P and Q are constants)

$$P - Q \cdot t_R = \log_{10}([\eta]_i M_i) = \log_{10} K_i = \log k + (1 + \alpha)\log_{10} M_i$$

These explain the linear form of the plot labelled as C in Fig. 2.6.

The highest yield of information appears to come when proton (^{1}H) NMR is used as the detection technique, particularly when the polymer has chiral centres. Figure 2.7 shows profiles of signal strength (proportional to the amount of polymer) versus chemical shift (δ) at a series of elution times (t_R) for a sample of fully isotactic poly(methylmethacrylate) (PMMA) in deuterated chloroform (CDCl$_3$) (so that only the polymer gives rise to ^{1}H signals). Only a small volume (6.0×10^{-5} dm^3) of solution gave rise to signals when the elution rate was 2×10^{-4} dm^3 min^{-1}, so that good resolution in terms of the total elution time of the PMMA sample is achieved.

The obvious peaks are assigned to hydrogen atoms in a methoxy group

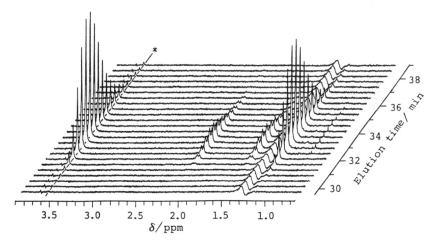

Fig. 2.7. A series of scans of chemical shift (δ) versus elution time generated using a gel permeation chromatography system with on-line ^{1}H (proton) NMR detection. The polymer is isotactic poly(methylmetacrylate) ($\bar{M}_n = 1.26 \times 10^4$) and the eluent was deuterated chloroform containing 0.5 per cent fully deuterated ethanol flowing at 0.2 cm^3 min^{-1}. The asterisk marks the chemical shift of the methoxy proton. Reproduced with permission of the Society of Polymer Science, Japan, from Hatada, K., Ute, K., Kashiyama, M., and Imanari, M. (1990). Direct determination of molecular weight and its distribution by the absolute calibration method using the on-line GPC/NMR. *Polymer Journal*, **22**, 218–22.

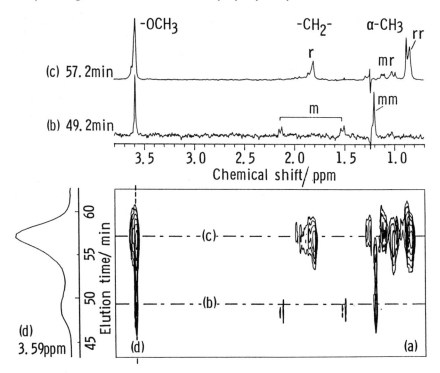

Fig. 2.8. The results of on-line detection by ^{1}H (proton) NMR of a not fully isotactic sample of poly(methylmethacrylate) eluted from a gel permeation chromatography column using deuterated chloroform as the eluent at a flow rate of 0.2 cm^3 min^{-1}. (a) A contour diagram of signal intensities as functions of chemical shift and elution time, (b) and (c) cross sections of signal intensities as the elution times indicated, (d) profile of the intensity of the signal corresponding to methoxy hydrogen atoms as a function of elution time. Reproduced with permission of Springer-Verlag, Heidelberg, from Hatada, K., Ute, K., Kitayama, T., Nishimura, T., Kashiyama, M., and Fujimoto, N. (1990). Studies of the tacticity of anionically prepared PMMAs by on-line GPC/NMR. *Polymer Bulletin*, **23**, 549–54.

$(-O-CH_3)$ at $\delta \equiv 3.6$ ppm, in a methylene link $(-CH_2-)$ at $\delta \equiv 2.1$ and 1.5 ppm and in a methyl group $(-CH_3)$ at $\delta \equiv 1.2$ ppm.

Figure 2.8 shows corresponding profiles for a much less stereoregular PMMA than that concerned in Fig. 2.7. Parts (b) and (c) are spectra showing peaks assigned to dyads (m and r) (Section 1.4) and triads (mm, mr, and rr) as sensed by the protons in the methylene and the *a*-methyl groups respectively of the $-CH_2-C(CH_3)(C(=O)OCH_3)-$ monomer residues, in which right-hand carbon atom of the backbone, to which the α-methyl group is directly attached, is chiral. The $-CH_2-$ units are thus sandwiched between pairs of chiral centres which bear one relatively polar group $(-(C=O)OCH_3)$ and thus the proton resonances in these methylene units are sensitive in terms of the resultant chemical shift to whether the adjacent chiral centres are the same (m) or different (r) in stereoconfiguration. In the former case (m dyad) the hydrogen atoms in the $-CH_2-$ units are in different microenvironments and thus give rise to two peaks in the NMR spectrum. The resonances of the hydrogen atoms of the $\alpha-CH_3$ groups are sensitive not only to the chiralities of the carbon atoms to which they are attached directly but also to those of the neighbouring two monomer residues.

Figure 2.8(a) is a contour diagram which is effectively the two-dimensional way of representing the corresponding three-dimensional form of Fig. 2.7. The locations of the sections across this diagram which corres-pond to the spectra labelled as (b) and (c) are shown. The peak originating from the methoxy group is labelled (d) within part (a) and the profile of its signal strength, proportional to the mass concentration of the PMMA, is shown as part (d) on the left of Fig. 2.8. What this figure reveals is that the lower molecular weight PMMA molecules tend to be syndiotactic (i.e. r dyads and rr triads predominant) whereas the heavier PMMA molecules are overwhelmingly isotactic (i.e. m and mm peaks predominantly). The breadth of the distribution of molecular masses and its bimodal nature is apparent in Figure 2.8(d) particularly.

Figure 2.9 illustrates the conversion of a scale of elution time into a scale of relative molecular mass (MW) for a third sample of PMMA. This sample is mainly isotactic but only the longest chains are purely so. Shorter chains have evidently significant atactic and/or syndiotactic contents.

The computer software which is supplied now with most GPC instru-

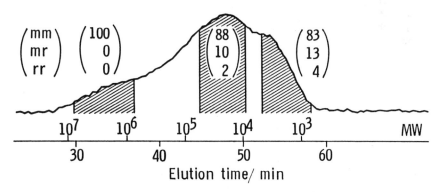

Fig. 2.9. Gel permeation chromatographic profile of the intensity of the ^{1}H (proton) NMR signal corresponding to hydrogen atoms in the methoxy ($-OCH_3$) group in a sam-ple of poly(methylmethacrylate) for which the percentages of the triads at three stages are shown. The horizontal scales show corresponding relative molecular weights (MW) and elution times. Reproduced with permission of Springer-Verlag, Heidelberg, from Hatada, K., Ute, K., Kitayama, T., Nishimura, T., Kashiyama, M., and Fujimoto, N. (1990). Studies on the molecular weight dependence of tacticity of anionically prepared PMMAs by on-line GPC/NMR. *Polymer Bulletin*, **23**, 549–54.

ments includes programs which generate values of $\bar{M}_n$, $\bar{M}_w$, and $\bar{M}_v$ for the polymer sample concerned from the signals taken in through the interface with the detector. Conventional practice is the main impetus for obtaining these averages. It is often advantageous to have dual detectors with different sensitivities to higher and lower parts of the molecular mass range. For example a differential refractometer is more sensitive in the lower part of the range because it responds to number rather than to size of the dissolved molecules. Low-angle laser light scattering (LALLS) responds to turbidity which is governed by the product of molecular size and concentration. LALLS is thus more sensitive to heavier molecules and is thus an obvious partner for a differential refractometer. This combination of detectors was used to generate the data of Table 2.2.

2.3 Liquid-phase osmometry

Osmometry is used reasonably widely to determine $\bar{M}_n$ and has been significant in this connection over a long time. Thus a brief section on this is appropriate here.

Osmotic pressure (π) is a colligative property of a solution and thus depends on molar concentration of the solute but not directly on the size of the molecules for a specified molar concentration. The underlying principle is that small molecules of solvent can pass through the pores of the semipermeable membrane which separates the solution of the polymer from the pure solvent. But the membrane is impenetrable by the molecules of the polymer on account of their size. If allowed to, solvent will pass through the membrane from the pure solvent to give rise to increased hydrostatic pressure associated with the rise of the liquid meniscus in a tube emerging vertically from the compartment containing the solution. This is the basis of the *static equilibrium method* in which π is evaluated from the final difference in height (h) between the liquid levels in the tubes from both compartments as $\pi = g \cdot \rho \cdot h$, g being the gravitational constant and ρ the solution density. The main disadvantage of this straightforward procedure is the long period of time, often days, which is usually required for h to reach its limiting value.

Modern osmometry is based upon the *dynamic equilibrium method*. This works on the principle that solvent does not move through the membrane when the osmotic pressure is exactly balanced by equivalent counterpressure. Typically a cell of volume less than $0.001 \, dm^3$ made of stainless steel is divided by the membrane, commonly a modified form of cellulose. Part of the wall of the part of the cell filled with the solution will be a stainless steel diaphragm and, with pure solvent filling the other part of the cell, it is the strain applied to this to counterbalance the osmotic driving force which is measured. The instruments available nowadays incorporate sensitive electronics and can display the equivalent value of π within minutes at most. Maintenance and cleaning of this type of instrument is particularly easy.

Polymer solutions are very nonideal, as would be expected when the dissolved molecules are enormous in comparison with solvent molecules. The osmotic pressure is expressed quantitatively by a virial equation in terms of the mass concentration (c)

$$\pi/c = (RT/\bar{M}_n) + A_2 \cdot c + A_3 \cdot c^2 + \ldots . \qquad (2.5)$$

This indicates that the intercept on the π/c axis at $c = 0$ is equal to $RT/\bar{M}_n$. A_2 is termed the second virial coefficient and its value increases when the solvent and polymer molecules interact together more strongly, i.e. the solvent is better. Higher power terms are not of substantial significance in general in the present context and plots of π/c versus c are close to being linear, allowing relatively easy extrapolation to $c = 0$. Figure 2.10(b) shows a representative example of this type of plot, with the 'raw data' of corresponding values of the head (h) of acetone (solvent) expected in a static equilibrium system plotted against *molar* concentration of the polymer shown in Fig. 2.10(a); the values of the last parameter emphasize how dilute these solutions are in conventional terms. The dashed line in part (b) corresponds to the linear plot expected if A_3 was insignificant. The intercept is $(\pi/c)_{c \to 0} = 6.0 \, \text{J kg}^{-1}$, leading to the deduction that

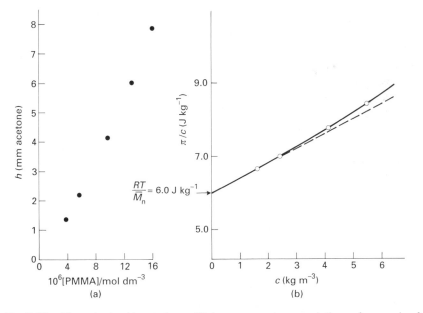

Fig. 2.10. Plots obtained by static equilibrium osmometry on solutions of a sample of poly(methylmethacrylate) (PMMA) in acetone at 303 K. (a) Plot of head *(h)* versus molar concentration of PMMA, (b) plot of π/c versus c, where π is osmotic pressure and c is mass concentration of PMMA.

$\bar{M}_{\mathrm{n}} = RT/6.0 = 8.314 \times 303/6.0 = 420 \, \mathrm{kg \, mol^{-1}}$ and hence $\bar{M}_{\mathrm{n}} = 4.20 \times 10^5$ in conventional relative terms.

Osmometry offers one considerable advantage over the viscometry and GPC methods in yielding an *absolute* result, i.e. it does not require calibration using standardized samples. Thus in view of the reasonably simple operational procedures of 'automatic high-speed membrane osmometers' now available commercially, it is not surprising that this remains a common means of obtaining information on the average masses of polymers. However no indication of the degree of polydispersity is provided in isolation.

Exercises

2.1. Match each of the items (i)–(ix) in the left-hand column with the most appropriate one from the right-hand column.

(i) $\bar{M}_{\mathrm{n}}$	(a) GPC universal calibration plot
(ii) $\bar{M}_{\mathrm{w}}$	(b) Mark–Houwink–Sakurada
(iii) $\bar{M}_{\mathrm{w}}/\bar{M}_{\mathrm{n}}$	equation
(iv) intrinsic viscosity	(c) rigid polymer molecule
(v) $\alpha < 0.6$	(d) $\Sigma N_i M_i^2 / \Sigma N_i M_i$
(vi) viscosity number	(e) $(\eta/\eta_0 - 1)$
(vii) $\alpha \sim 1.00$	(f) $\leqslant \bar{M}_{\mathrm{v}}$
(viii) $\log_{10} [\eta] M$ vs t_{R}	(g) ideal osmotic pressure
(ix) $RT/\bar{M}_{\mathrm{n}}$	(h) poor solvent
	(i) polydispersity index.

2.2. Evaluate $\bar{M}_{\mathrm{n}}$ and $\bar{M}_{\mathrm{w}}$ for a sample composed of monodisperse fractions, each composed of N_i molecules of relative molecular mass M_i, as specified below

N_i	2000	5000	7000	3000	1000
$10^{-5} M_i$	1.00	2.00	3.00	4.00	5.00

2.3. Evaluate $\bar{M}_{\mathrm{v}}$ for the data given in Exercise 2.2 above using $\alpha = 0.80$.

2.4. Will any or all of the values of $\bar{M}_{\mathrm{n}}$, $\bar{M}_{\mathrm{v}}$, and $\bar{M}_{\mathrm{w}}$ measured for a particular polymer sample be expected to depend in general upon the identity of the solvent used?

2.5. The following values of intrinsic viscosity $[\eta]$ of standardized samples of polystyrene having very narrow ranges of relative molecular mass centred on M have been measured for solutions in a mixed chloroform/hexafluoroisopropanol (98:2 v/v) solvent at 298 K

$[\eta] (\mathrm{dm^3 \, kg^{-1}})$	17.3	33.9	67.9	165
$10^{-4} M$	1.85	4.89	13.3	48.0

Devise the corresponding Mark–Houwink–Sakurada equation. *Source*: Weisskopf, K. (1988). *Journal of Polymer Science*, **A26**, 1919–35.

2.6. Polyoxymethylene $((-O-CH_2)_n-)$ in solution in dimethylformamide (DMF) at 408 K has intrinsic viscosities $([\eta])$ which are expressed by eqn (2.4) with the constants having values of $10^3 k \, (\mathrm{dm^3 \, kg^{-1}}) = 9.32$ and $\alpha = 0.7895$.

The corresponding universal calibration plot for gel permeation chromatography (GPC) was expressed by $\log_{10}([\eta]M)\,(\mathrm{dm^3\,km^{-1}}) = 17.30 - 702.3$ $(V)\,(\mathrm{dm^3})$ when the flowrate of the DMF solvent was $1.00 \times 10^{-3}\,\mathrm{dm^3}$ $\mathrm{min^{-1}}$ at 408 K. What is the relative molecular mass of polyoxymethylene which is emerging from the GPC column when the volume of eluent (V) is $1.500 \times 10^{-2}\,\mathrm{dm^3}$?

Source: Ogawa, T. (1990), Journal of Liquid Chromatography, **13**, 51–61.

2.7. Samples of linear poly(ethylene imine) $((-\mathrm{NH-CH_2-CH_2})_n-)$ with very small ranges of relative molecular masses centred on M were dissolved in methanol at 298 K. Viscometry of the resultant solutions gave the following set of gradients (G) for the apparently linear plots of specific viscosity vs mass concentration for different M

$G\,(\mathrm{dm^3\,kg^{-1}})$	21.1	76.7	137	172
$10^{-3}M$	3.00	11.7	21.5	27.5

What may be deduced about the nature of the polymer molecule in solution in methanol?

Source: Weyts, K. F., W. M. Goethals, Bunge, W. M., and Bloys van Treslong, C. J. (1990). European Polymer Journal, **26**, 445–7.

2.8. The concept of universal calibration for gel permeation chromatographic analysis of relative molecular masses of polymers was proposed by Grubisic, Z., Rempp, R., and Benoit, H. (1967). Journal of Polymer Science, **B5**, 753. If the subscript PS refers to the approximately monodisperse (standardized) sample of polystyrene and the subscript X refers to the fraction of a polydisperse polymer which emerges with the same eluted volume of solvent from the same gel permeation chromatography system, show that the following equation is justified when universal calibration applies

$$\log_{10} M_X = (1 + \alpha_X)^{-1} \cdot \log_{10}(k_{PS}/k_X)$$
$$+ \{(1 + \alpha_{PS})/(1 + \alpha_X)\} \cdot \log_{10} M_{PS}$$

where M_{PS} and M_X are relative molecular masses and k_{PS} and k_X and α_{PS} and α_X are the constants in the corresponding forms of the Mark–Houwink–Sakurada equations.

2.9. A sample of poly(ethylene terephthalate) (PET) has been dissolved in a mixed solvent composed of chloroform and hexafluoroisopropanol (98:2 v/v) to give a series of solutions of different mass concentrations (c). An osmometer was used to measure the corresponding osmotic pressures (π). Given that a plot of π/c vs c gave an intercept equivalent to $94.2\,\mathrm{J\,kg^{-1}}$ and the point specified by $\pi/c = 141\,\mathrm{J\,kg^{-1}}$ and $c = 8.70 \times 10^{-3}\,\mathrm{kg\,dm^{-3}}$ lies on the plot, evaluate (a) the second virial coefficient (A_2) and (b) the average number of constitutional repeat units per PET molecule.

Source: as for Exercise 2.5.

3
Microscopic features of bulk polymers

3.1 General aspects

It might be thought at first that a solid composed of very long molecular chains would simply have a tangled, spaghetti-like structure at the microscopic level. The real picture can be much more interesting and gives rise to the special properties associated with polymeric substances.

The obvious first question is 'how long do the individual molecular chains have to be before the bulk material shows the special properties of polymers?'. Before any answer can be given, there is a need to consider what happens to long chains within continuous bulk phases. The molten (liquid) state is a good starting point because its microscopic structure can only resemble a mass of live worms, all entangled together and wriggling about. Expressed more scientifically, in the melt each chain is very loose as regards rotations around the bonds in the backbone so that segments of it indulge in rapid twisting motions. The chains are able to slip past one another in general but tend to tangle: the result at the macroscopic level is a viscous liquid rather like a syrup.

Consider now what may happen as an initially molten polymer is cooled down gradually. When the temperature has come down far enough, the solid phase which has the strongest attractions between individual chains would be expected to appear first. In the reverse melting process this phase is the last bit of the solid to disappear on heating, i.e. the thermally most stable part. Attraction between the chains is particularly promoted when there are most individual points of mutual cohesion along their lengths. It is obvious that this requires that the chains are highly ordered in a regular close-packed array, features which are associated with the general phenomenon of *crystallinity*. Thus the solid phase which comes out of the molten polymer initially is expected to have crystalline characteristics with segments of chains aligned with one another as much as possible. Small zones in which the chains are aligned thus appear at various locations within the amorphous mass of molten polymer. Figure 3.1 shows a microscopic section in which there are three separate zones (labelled A, B, and C) in which crystallization has commenced in which the chains are represented as parallel lines. Outside these zones enclosed by dashed lines, the polymer

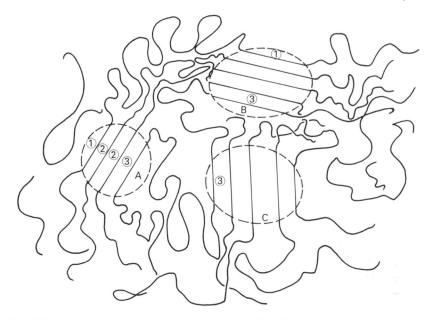

Fig. 3.1. Representation of microscopic structure in a thermoplastic polymer sample. Each line corresponds to the backbone of a polymer chain and parallel lines (each numbered to indicate the particular chain of which they are part) encircled by dashed lines represent zones of crystallinity (crystallites) (labelled as A, B, and C).

is amorphous and indicated by 'spaghetti'. The chains of interest are labelled 1, 2, and 3. What is apparent immediately is that the same molecular chain may be involved in more than one crystalline zone; for example chain 1 appears in A and B whilst chain 3 appears in A, B, and C. Moreover one chain may be incorporated into the same crystalline zone at different points along its length; for example chain 2 appears twice in zone A. Consider what will happen on cooling a little more. There is a driving force for the expansion of the crystalline zones, which represent the greatest thermodynamic stability. But obviously that there is a limit to how far this process can go. Eventually an energy impasse is reached, when in order to extend one crystalline zone further, others in the vicinity must be broken up to extricate the required segments of chains from them. The energy demand for this is large and is not available at the limit of growth of crystalline zones. Once formed in a cooling bulk polymer, crystalline zones (*crystallites*) will not be inclined to disintegrate to allow segments of chains to be extracted for assemblage of larger crystalline regions. Crystallite dimensions are $10\,\mu$m typically. So even in the cases of the polymer chains which are most amenable to crystallization, the formation of large crystals (say visible to the naked eye) can be regarded as not possible. The crystallites in a solidified polymer are separated by a continuous amorphous phase composed of tangled chains. The term *microcrystalline* is an appropriate description of bulk polymers composed of crystallites embedded in amorphous material. In polymers which have been carefully annealed at temperatures close to that at which crystallites first appear on cooling from the melt, the degree of (micro) crystallinity can extend up to 90 per cent or so.

At the basic level, the difference between flexible and rubbery (elastomeric) polymeric materials arises from the significant degrees of crystallinity of the former as opposed to the almost totally amorphous natures of the latter. The common thermoplastics, such as nylon 6,6, PET, and polyethylene, are flexible at ambient temperatures; usually 40–70 per cent of their bulk is made up by crystallites, which are rigid because of the immobility of the incorporated chain segments. The key microscopic process which occurs throughout an elastomeric polymer is chain mobility via *segmental rotation*, which allows the material to give way easily under applied stress. The amorphous microstructure allows segments of chains to pivot around via rotations about bonds in their backbones to move into the gaps (void volumes) between neighbouring chains within the jumbled assembly. Then a segment of another chain can twist around to move into the vacant space left by the first and so on. A concerted set of such segmental rotations relieves strain created by an applied stress, for example a pull. In reverse, on removal of the applied stress, the mobility of segments allows them to 'unwind' rapidly, restoring the original shape of the bulk elastomer, i.e. it 'snaps' back into shape after deformation. The slippage of chains past one another is another way of relieving strain but this is undesirable in useful elastomers since the final result would be disintegration of the material, as happens when chewing gum is stretched. Accordingly 'rubbers' have crosslinks so that neighbouring chains are joined together at a few points (commonly by chemical bonds), allowing most chain segments to be mobile whilst stopping the bulk from falling apart when stretched (say).

A flexible thermoplastic is not rubbery because only chain segments in its amorphous regions are mobile. The rigidity of the embedded crystallites reinforces the overall structure to an extent which prevents sufficient chain segments being mobile to produce elasticity.

It is now apparent in general terms how long molecular chains have to be to give the special properties, most obviously elasticity and flexibility, of polymeric substances. The chain for an elastomer must be sufficiently long for small parts of it to be effectively immobilized by crosslinking without much effect on the overall mobility via rotations about bonds in the backbone. On the other hand, the chain of a thermoplastic must be long enough to allow several segments to be incorporated into different crystallites, with intervening segments in an amorphous phase. Obviously it is impossible to specify a universal minimum number of monomer residues (or CRUs) for the appearance of these properties in the corresponding bulk material, if only because it will vary from polymer to polymer. But it is easy to see that ten monomer residues will not be enough and that around 50 would be the lowest possibility. Look for instances at the $\bar{M}_n$ values for commercial polymers in Table 2.2; the lowest correspond to 82 monomer residues for polystyrene and 357 for polyethylene.

Rigid polymeric solids have immobile chains, even if the degree of crystallinity is insignificant and the bulk is wholly amorphous. There is a critical temperature for a particular polymer below which segments of chains no longer have the ability to pivot around, when it may be said that rotations about bonds in the backbone have 'frozen up'. Both elastomers and thermoplastics become *glassy* and thus hard in this way if they are cooled sufficiently. Thus the temperature which marks the change from rubbery or flexible solids respectively to glassy forms is known as the *glass transition temperature* and is denoted as T_g accordingly.

This section leads into the main topics of this chapter. These are how the structures of chains affect microcrystallinity, segmental rotation, and associated properties and also how these may be investigated experimentally.

3.2 Microcrystallinity and drawing

Consider polyethylene, the synthetic polymer with the simplest structure $(-(CH_2-CH_2-)_n-)$. The most ordered arrangement of this chain has

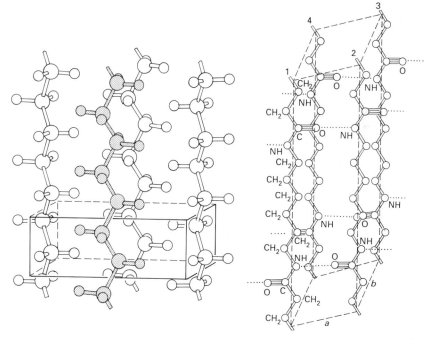

Fig. 3.2. The arrangement of the component atoms and bonds in the polymer chains in crystallites in polyethylene (*left*) and nylon 6,6 (*right*). The boxes enclose the unit cells. Reproduced by permission of the Royal Society of Chemistry from Bunn, C. (1975), Macromolecules—the X-ray contribution, *Chemistry in Britain*, **11**, 171–7.

all of the carbon atoms in one plane, when the C—C bonds form a zigzag. This is also the arrangement which produces *maximum extension* (or maximum end-to-end length) of the chain. These zigzag sections of chains pack together closely in the crystallite, as represented in Fig. 3.2(a). It is the regularity of these chain sections which allows them to pack together very closely and this is the most important single factor promoting microcrystallinity in the corresponding bulk polymer. Despite the fact that the interchain forces, van der Waals in type, are relatively weak in polyethylene, samples of this can be highly crystalline. In some cases hydrogen bonds provide stronger interchain attractions, nylon 6,6 being the obvious example with linkages represented as >C=O———H—N< between chains in which the central feature is an O———H hydrogen bond. These types of linkages are shown in Fig. 3.2(b) which represents the arrangement of sections of four chains in a crystallite of nylon 6,6. PET (poly(ethylene terephthalate)) has a highly regular chain structure but has no potential for hydrogen bonding between chains. Nevertheless PET can be a highly crystalline polymer which may be interpreted in terms of the conjugated systems of bonds represented as below which ensures that all of these atoms lie with one plane, a feature which can be expected to allow the chains to pack together closely in a regular array.

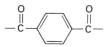

The 'melting' of a thermoplastic does not have the usual significance applied to normal chemical compounds. The bulk polymer usually softens over a fairly wide range of rising temperature, often tens of degrees, until at a particular temperature denoted as T_m a viscous liquid phase is achieved. T_m may be regarded as the highest temperature at which any solid material can exist within an otherwise molten polymer and is often referred to as the 'melting temperature'. These remnants must be the most stable parts of the solid phase, which are inevitably within larger crystallites in the initial thermoplastic at lower temperatures.

Consider what will happen when a sample of a thermoplastic with a regular chain amenable to crystallization is stretched when the temperature is between T_g and T_m. This process is termed *drawing* and with certain polymers it results in a thin fibre with greatly improved mechanical properties. In the course of drawing, the polymer sample, usually commencing in the form of a thin rod, does not gradually become uniformly thinner. The behaviour is as represented in Fig. 3.3. The rod thins ('necks down') at one location at first. Continued drawing makes the drawn section take in more

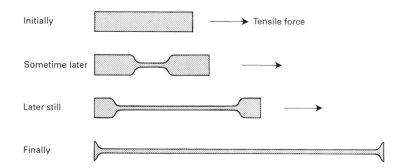

Initially

Sometime later

Later still

Finally

Tensile force

Fig. 3.3. Representations of the shapes at various stages during the drawing of a thin rod into a fibre.

of the polymer, until finally virtually all of it is in the resultant fibre. What is extraordinary is that the cross-sectional dimensions of the drawn and undrawn sections hardly change as drawing proceeds and it seems as if the fibre is being pulled out of the residual parts of the original rod. The *draw ratio* is defined as the final length divided by the original length of the same amount of polymer and is often in the range four to six.

In an undrawn polymer, crystallites are likely to be randomly orientated with respect to one another (Fig. 3.1). The chains react to the directional stress of drawing by reorientations which tend to result in more of their lengths lying in this direction. Drawing thus tends to produce greater alignment of chains overall, allowing crystallinity to be enhanced with the sections of chains in the crystallites tending to align with the fibre length. When the starting material has relatively low crystallinity, drawing can be expected to increase this. But when the degree of crystallinity in the *isotropic* mass which is drawn is relatively high, the resulting fibre may not have a significantly larger degree of crystallinity as such, but it will be *anisotropic* (i.e. shows different properties in different directions) because the crystallites are much more aligned. In this case many of the pre-existing, more randomly orientated zones of microcrystallinity must be destroyed during the drawing to be succeeded by newly formed microcrystalline zones which tend to have their component chain sections directed along the fibre (see Fig. 4.10 to follow).

Polymers which are synthesized for fibre production will usually be spun directly from the bottom of the reactor (usually an autoclave). When the polymer is like nylon 6,6 in not decomposing significantly when it is at a temperature just above T_m, the melt is extruded through a spinneret which has 100–1000 fine holes typically. The resulting threads cool rapidly and are thus mainly amorphous. Drawing of these threads induces high degrees of crystallinity in zones which reinforce the overall structure and thus drawing is essential to produce the very strong fibres (Section 1.2). This procedure is termed the *melt-spinning technique* and it is applied generally to polyamides, polyester, and polyalkenes for fibre production.

Both the length of the molecular chains and the value of the draw ratio applied are important for the use of the polymer. Longer chains will be expected to stiffen the bulk polymer; not only are they more entangled in amorphous regions but each chain is likely to be involved in more crystallites than would be the case for shorter chains. Thus the values of $\bar{M}_n$ of PET for use as an engineering thermoplastic (for example for making parts of vehicles, etc.) are significantly higher (typically in the range $3–5 \times 10^4$) than in PET intended for fibre production ($1–1.5 \times 10^4$). In the latter case, lower draw ratios result in the more flexible and softer fibres which are suitable for weaving into clothing. Higher draw ratios are used to produce the tougher and harder PET fibres which are used to make carcasses of vehicle tyres for instance, when much of the robustness reflects the higher degrees of orientated crystallinity and consequent greater reinforcement of the stiffness and strength of the more drawn fibres.

A key point is that the interchain attractions after drawing are increased above the point at which there is any tendency for the material to return to its undrawn state.

3.3 Amorphous polymers

There are several properties which are dependent on the absence of microcrystallinity to any extent in a bulk polymer. Amongst these are elasticity and optical transparency. Some polymers like polyacrylonitrile are almost totally amorphous and remain so after drawing out into a useful fibre.

The main feature of an elastomer (or 'rubber') is its ability to recover rapidly on removal of stress; an instance is the snapping back of a stretched elastic band on release. The discussion in the last section should make it evident that useful elastomers must have glass transition temperatures (T_g) well below ambient because rapid segmental rotation is the underlying phenomenon at the microscopic level. Consider the stretching of an elastomer in comparison with the drawing of a fibre. In both instances the chains must rearrange their lengths into the stretching direction to a greater extent. In an elastomer the virtual absence of crystallites before stretching is very important to allow almost all segments to rotate easily so that the chains can tend to align in the direction of extension. After stretching it is important that interchain attractions are not strong enough to confer thermodynamic stability on the extended elastomer. Although it is an amorphous polymer, commercial polyacrylonitrile ($-(CH_2-CH(CN)-)_n-$) is drawn to give one of the important synthetic fibres which is virtually devoid of crystallites. The key to explaining the nonelastomeric nature of poly-acrylonitrile is the strong electron-withdrawing ability of the cyano ($-C\equiv N$) group which gives rise to a negative charge on this pendant group with a counterbalancing positive charge located on the backbone, i.e. a dipole. Commercial polyacrylonitrile is atactic (Section 1.4) and it seems that the random orientation of the $-C\equiv N$ groups at successive chiral centres along the chains gives sufficient irregularity to prevent the close packing needed for crystallinity. Nevertheless the strong interchain attraction which results from the large number of charged sites on one chain interacting with oppositely charged sites on neighbouring chains gives polyacrylonitrile the greatest cohesive energy density (the energy required to abstract a molecular chain from the solid) of any of the common polymers; poly(ethylene terephthalate) and polyethylene have roughly half and quarter respectively of the cohesive energy density of polyacrylonitrile, even if they are substantially crystalline. Drawing of polyacrylonitrile yields a strong fibre in which there is considerable alignment of the molecular chains along its length but no significant degree of crystallinity. Thus it may be perceived that it is the alignment of chains in a permanent manner which is important for strong fibres, a point which is developed further in Section 9.3: microcrystallinity is not essential in this respect, simply being one way of achieving this with suitable polymers.

Elastomers have relatively small interchain interactions of necessity. When an elastomer is fully stretched, in the sense that further extension would involve rupture of chemical bonds in the backbone, its molecular chains will have been substantially aligned in the direction of stretch. In order to ensure complete recovery from such large deformations, some interlinking of chains is essential to give a permanent three-dimensional network structure but of a very loose nature. What is required is some points of strong linkage between each chain and its neighbours, sufficient to stop slippage but not enough to interfere with the mobility of most segments of the chains. Commonly these links are chemical bonds, but as will be found in connection with thermoplastic elastomers (Section 8.4), there are other ways in which chain slippage can be prevented in polymers with rubbery properties.

The common elastomers are derived from monomers possessing two double bonds, represented by the general formula $CH_2=CH-C(X)=CH_2$ and giving rise to a CRU with one double bond represented as $-CH_2-CH=C(X)-CH_2-$. It is the presence of double bonds in the chains which allows the crosslinks to be created subsequently, in a process known generally as *vulcanization*. Sulphur was the original vulcanizing agent (Section 1.2), giving rise to links between chains composed of two or three sulphur atoms usually. A representative pair of crosslinks between chains in vulcanized natural rubber (polyisoprene, $X = CH_3$) is shown in Fig. 3.4(a); these have effectively used the electron density of the π

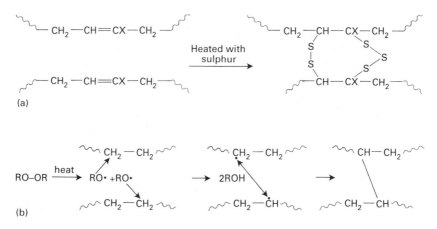

(a)

(b)

Fig. 3.4. Representations of typical processes of the formation of crosslinks using (a) sulphur (vulcanization) and (b) an organic peroxide (ROOR).

components of the double bonds which existed prior to vulcanization to make the bonds to the chains. The extent of incorporation of sulphur must be limited and some 3 per cent by weight is found generally to produce a useful elastomer. This corresponds to one sulphur atom per 16 CRUs in polyisoprene, suggesting that the average length of segment between crosslinks is equivalent to at least 16 CRUs. Organic peroxides, ROOR generally, are other agents which are used to introduce crosslinks into synthetic polymers. In the course of thermal decomposition after mixing into the bulk polymer, these generate alkoxy (RO·) radicals which abstract hydrogen atoms from sites such as $-CH_2-$ in the chains leaving radical sites $-C(\cdot)H-$ which combine with others on different chains to create crosslinks. On this basis one crosslink should be created per molecule of ROOR and this is often so. Figure 3.4(b) represents this mechanism.

A key point here is that polymers with saturated chains can be crosslinked, for example polyethylene, using say, ditertiarybutyl peroxide $((CH_3)_3C-O-O-C(CH_3)_3)$. Not all vinyl polymers are suitable, some such as polypropylene and poly(vinyl chloride) tending to be degraded rather than crosslinked in the presence of organic peroxides.

When a polymer is irradiated with high energy electrons, obtained from a van de Graff generator typically, radical sites are created on the chains which form crosslinks when those on different chains combine. For example, poly(vinyl chloride) can be crosslinked in this way, giving a bulk polymer which is more resistant to deformation, penetration by solvents, and softening on heating, features which are expected to be conferred by crosslinking. Some polymers, notably high-density polyethylenes, are difficult to crosslink by means other than using high-energy radiation.

The other property of bulk polymers of interest here is optical transparency when the sample is physically thick and mechanically strong. The main feature required for this is that the material must not be composed at the microscopic level of phases of significantly different refractive indices. Crystallites will usually not be the same as the amorphous polymer in this respect and thus act to scatter light as it passes through a crystalline polymer making it appear translucent at the macroscopic level, as with usual samples of polyethylene. Transparency in bulk is only to be expected when the polymer sample concerned is wholly amorphous. Moreover when it is to serve as an optical glass, the polymer must be rigid, which demands that its glass transition temperature is well above ambient. The thermoplastics which are commonly used as optical glasses are poly(methylmethacrylate) (PMMA) and polystyrene. PMMA is commonly known by trade names of Perspex (ICI) or Lucite (Du Pont) and finds wide usage as windows in vehicles. Some stereoconfigurations of PMMA have been specified in connection with Figs 2.7–2.9. The commercial form of PMMA is typically 70–75 per cent syndiotactic with the remainder largely atactic. This level of stereoregularity is insufficient to allow any significant formation of crystallites and thus normal forms of PMMA are amorphous and transparent. The glass transition temperature of 378 K ensures the rigidity required for a window material.

Similar considerations apply for polystyrene, commercially available in its atactic form, which has outstanding optical clarity and the highest refractive index (1.60) of any of the common polymers. The glass transition temperature (373 K) is about the same as for PMMA. But the common polystyrenes are less resistant to the effects of weather and ultraviolet light and also have lower impact strengths than common PMMAs.

3.4 Glass transition temperatures (T_g)

It is apparent from remarks made in previous sections that any structural feature which hinders the rotation of segments of the molecular chain should increase the value of T_g of the bulk polymer. This section explores the specific factors which exert an influence on this phenomenon which is manifested as a transition from a hard glassy solid to a rubbery (for elastomers) or flexible (for thermoplastics) solid on warming through T_g.

The simple molecule n-butane ($CH_3CH_2CH_2CH_3$) provides a useful starting point. *Internal rotation* about the central $C—C$ bond involves passing over energy barriers. *Conformations* are defined as a series of states achieved through internal rotation. Consider the two conformations of n-butane which are represented in Fig. 3.5(a). The 'staggered' conformation places the two bulkiest substituents, the methyl (CH_3) groups, as far apart as possible; this is the conformation, known as '*trans*', with least steric interference and hence lowest potential energy accordingly. The 'aligned'

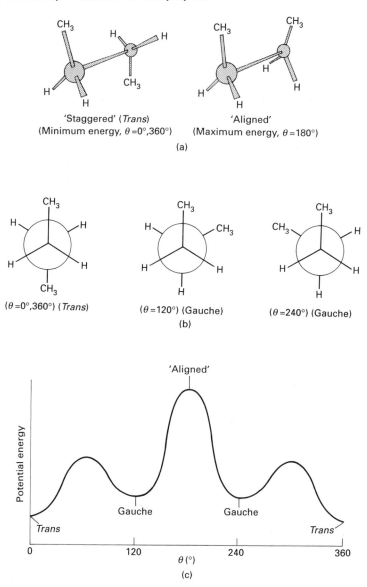

Fig. 3.5. Representation of conformations of *n*-butane and their energies. (a) Staggered (*trans*) and aligned conformations, (b) Newman projections of *trans* and *gauche* conformations, (c) profile representing the general form of the variation of potential energy with angle of rotation (*θ*) about the central C—C bond.

conformation on the other hand must have the highest potential energy because it brings the methyl groups closest together. Newman projections of local minimum energy conformations are shown in Fig. 3.5(b). In addition to the *trans* conformation already mentioned, there are two others known as 'gauche' conformations, of higher potential energy than the *trans* because they bring the methyl groups closer together. Figure 3.5(c) is a plot of potential energy versus angle of rotation (θ) about the central C—C bond. The potential energy scale is indicated by estimates that the energy differences between 'aligned' and 'gauche' and between 'gauche' and '*trans*' are equivalent to around 13 kJ mol^{-1} and 3.3 kJ mol^{-1} respectively.

The equivalent representations for internal rotations about a single C—C bond within the backbone of polyethylene come from simply replacing the methyl groups with polymeric chains ($-(CH_2)_{2n}-$). It turns out that the energy barriers are relatively little changed by this substitution. *Segmental rotation* can occur, where a length of the chain moves via internal rotations within two or more of the component backbone bonds. Segments can indulge in rapid rotations at temperatures at which the average thermal energy in each bond is considerably larger than any energy barrier to internal rotation, such as that between gauche and aligned conformations. Figure 3.6 represents a simple case where a segment of eight bonds responds to an applied stress by internal rotations at the two ends (effectively pivoting) so that it moves into the void volume between this and the neighbouring chain.

It is the kinetics of internal rotations in individual bonds which is the key to understanding the glass transition phenomenon. The rate at which a system can pass over an energy barrier designated as $\Delta \epsilon$ (referring to one internal rotation) will be expected to be governed by the usual exponential factor $\exp(-\Delta \epsilon / kT)$, where k is the Boltzmann constant. $\Delta \epsilon$ might be viewed here as a measure of the potential energy barrier to internal rotation equivalent to that apparent in Fig. 3.5(c). Now consider a large assembly of polymer chains in a rubbery solid, such as an elastomer well above and T_g, and how it responds to an applied stress. Repeated microscopic actions of the type represented in Fig. 3.6 allow easy deformation, when $\exp(-\Delta \epsilon / kT)$ is large enough to correspond to extremely rapid internal rotation in backbone bonds. Most chain segments will only be able to rotate away from stress if at least one of the neighbouring segments has already rotated away from it to leave a suitable void volume for this segment to move into. Thus it is obvious that a large number of segments must rotate in sequence and very rapidly to give rise to rubbery properties at the macroscopic level. There is an analogy to a set of partially interlocking gear wheels which is helpful to understanding. In this all wheels rotate easily when the analogy is to an elastomer above T_g, so that a stress which is applied as torque on one wheel is relieved by the rotation of other wheels in the assembly.

Now consider the effect of lowering the temperature towards T_g. The rates of internal rotation in the polymer chains become less rapid as $\exp(-\Delta\epsilon/kT)$ becomes smaller. Indeed because the energy is statistically distributed, some individual internal rotations will have momentarily 'seized up', having at that instant insufficient thermal energy to overcome the energy barrier $\Delta\epsilon$. In the analogous gear wheel assembly, the equivalent circumstance is that some of the wheels have become 'locked', i.e. are unable to rotate. The system can tolerate a certain fraction of seized wheels, because it is only partially interlocked. But it is evident that there will be a critical value of this fraction, above which the whole assembly jams up with wheels locking their neighbours and so on. Returning to the polymeric system, when the equivalent critical fraction of segments loses the ability to rotate easily, the rest are frozen since the cooperative movement of segments required for rubbery properties is no longer possible. It is the huge number of individual bonds and segments within a bulk sample of polymer which ensures that the transition to a hard solid with immobile chain segments takes place at a sharply defined temperature (T_g) when $\exp(-\Delta\epsilon/kT)$ corresponds to the critical value in statistical terms for the kinetics of internal rotation.

Any factor which hinders internal rotation about bonds in the backbone will increase T_g in the corresponding amorphous polymer. Polyethylene

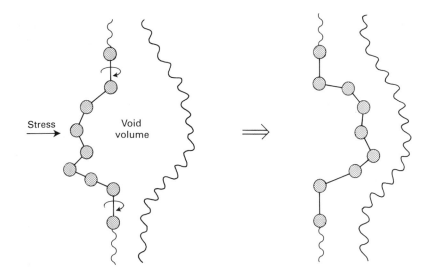

Fig. 3.6. Simple representation of the segmental rotation of a polymer chain (only backbone atoms and bonds shown) into an adjacent void volume under the influence of applied stress, when internal rotation takes place only about two bonds (hooped arrows).

may be taken as the basis of comparison for a series of thermoplastics of which CRU structures and values of T_g are shown in Table 3.1. The expected trend is seen for vinyl polymers of general CRU formula $-CH_2-CH(X)-$ since T_g rises as X becomes more bulky (along the series $X = H$, CH_3, C_6H_5) or as X becomes more polar (and bulkier) ($X = H$, Cl, CN). The insertion of an oxygen atom between CH_2 groups in the backbone would be expected to ease internal rotation and thus to lower T_g: this is seen along the series polyethylene, poly(ethylene oxide), poly(oxymethylene). When the CRU has a general structure represented as $-CH_2-C(CH_3)(X)-$, T_g is higher than for the corresponding amorphous polymer without the substituent methyl group, as is apparent with polystyrene compared to poly(α-methylstyrene) and with poly(methylacrylate) compared to poly(methylmethacrylate). The methyl groups in the latter polymers cause significantly more steric hindrance to internal rotations than do hydrogen atoms seemingly. The case of PET is interesting, when it is considered that a substantial part of the CRU $(C(=O)-C_6H_4-C(=O))$ is rigidly planar. But the remainder of the CRU $(-O-CH_2-CH_2-O-)$ would be expected to have relatively easy internal rotations by analogy with poly(ethylene oxide). The compromise between these extremes is reflected by a moderately high value of T_g for PET.

Hydrogen bonding raises T_g values, as would be expected; interchain linkages will evidently hinder segmental rotations in their vicinity. This is apparent in the relatively high T_g which for nylon 6,6 is just over double that for the polyester with a comparable chain but no potential for hydrogen bonding i.e. $(-C(=O)-(CH_2)_4-C(=O)-O-(CH_2)_4-O)_n-$ for which T_g is 155 K.

Table 3.1 Typical T_g values of common thermoplastics (atactic forms)

Polymer	CRU structure	T_g (K)
Poly(oxymethylene)	$-CH_2-O-$	198
Polyethylene	$-CH_2-CH_2-$	253
Poly(ethylene oxide)	$-CH_2-CH_2-O-$	206
Polypropylene	$-CH_2-CH(CH_3)-$	267
Poly(propylene oxide)	$-CH_2-CH(CH_3)-O-$	226
Polystyrene	$-CH_2-CH(C_6H_5)-$	381
Poly(α-methylstyrene)	$-CH_2-C(CH_3)(C_6H_5)-$	445
Poly(methylacrylate)	$-CH_2-CH(C(=O)OCH_3)-$	281
Poly(methylmethacrylate)	$-CH_2-C(CH_3)(C(=O)OCH_3)-$	394
Poly(vinyl chloride)	$-CH_2-CH(Cl)-$	354
Polyacrylonitrile	$-CH_2-CH(-C\equiv N)-$	378
Poly(ethylene terephthalate)	$-O-CH_2-CH_2-O-C(=O)-C_6H_4-C(=O)-$	340
Nylon 6,6	$-C(=O)-(CH_2)_4-C(=O)-NH-(CH_2)_6-NH-$	357

The polymers for which low values of T_g are essential are the so-called rubbers or elastomers. Commonly these have monomer residues (and CRUs) of general formula $-CH_2-CH=C(X)-CH_2-$ and internal rotations about $C-C$ bonds in the chains are expected to be relatively easy. It is important that all of the carbon atoms (other than any in X) of the monomer become part of the polymer chain backbone (i.e. 1,4-bonding), which makes the $-CH=C(X)-$ part planar where the angle subtended

by the $\begin{smallmatrix} H \\ \diagdown \\ C \diagup \end{smallmatrix} C=$ units is $\sim 120°$. The barrier to internal rotation about this

$C-C$ bond is in fact lower even than that in ethane, because only one of the bonds extending from the tetrahedral carbon atom can be aligned with either $C-H$ or $C=C$ at one time, as compared to the simultaneous alignment of three pairs of $C-H$ bonds in the highest energy conformation of ethane. Thus internal rotation about a $C-C$ bond in a polymeric chain is greatly eased when the next bond along the backbone is $C=C$, so that the polymer with CRU represented as $-CH_2-CH= C(X)-CH_2-$ may even be expected to have a lower T_g than that composed of $-CH_2-CH_2-CH_2-CH_2-$ units, i.e. polyethylene unless X is relatively large and/or polar. This point is borne out in Table 3.2.

Polysiloxanes are known commonly as 'silicones' and have remarkable properties, including chemical inertness and very low values of T_g. Figure 3.7 compares the geometries of the extended chain structures of isotactic polypropylene ($T_g = 255$ K) and one of the common silicones, poly (dimethyl siloxane) with T_g just over 100 K lower. It is the relatively large separation of the backbone silicon atoms and their substituents which accounts for the ease of internal rotations and hence the low T_g, which is even over 40 K lower than that of poly(oxymethylene) (Table 3.1).

3.5 Melting temperatures (T_m)

T_m refers to a phase transition which converts crystalline solid to liquid polymer. It is logical to denote the corresponding enthalpy and entropy changes as ΔH_m and ΔS_m. The corresponding Gibbs energy change (ΔG_m) is zero when the process is conducted under equilibrium conditions so that

Table 3.2 Typical values of T_g of common elastomers (rubbers)

Polymer	X	T_g (K)
Poly(cis-1,4-butadiene)	H	165
Poly(cis-1,4-isoprene)	CH$_3$	206
Neoprene	Cl	253

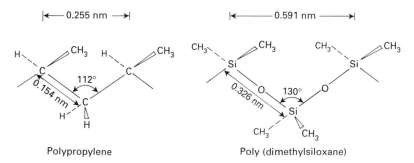

Fig. 3.7. Comparison of the geometries of the planar zigzag conformations of polypropylene and poly(dimethylsiloxane).

$\Delta H_m = T_m \Delta S_m$ and $T_m = \Delta H_m / \Delta S_m$. Table 3.3 shows representative values of these parameters for various common thermoplastics in which substantial fractions of bulk samples are usually crystalline. The 'mol' used here refers to one mole (i.e. 6×10^{23} as a number) of CRUs.

The first point of main interest is the large difference in the T_m values of polyethylene and poly(tetrafluoroethylene) (PTFE). The much lower T_m of polyethylene prohibits its use for coating internal surfaces of cooking utensils, as is a common usage of PTFE. What can be perceived from the entries in Table 3.3 is that it is the much lower value of ΔS_m for PTFE compared to that for polyethylene which is mainly responsible for the higher thermal stability of PTFE crystallites. Figure 3.8 shows a photograph of space-filling models of these polymer chains. What is obvious is that the fluorine atoms, much larger than hydrogen atoms, form an interlocking sheath along the PTFE chain. This difference is of particular significance for the molten phases, in which PTFE chains will be much stiffer than polyethylene chains, the latter being able to undergo almost free internal rotations to explore a much larger range of conformations. Suppose that it was possible to take a snapshot of an individual chain

Table 3.3 Values of T_m, ΔH_m, and ΔS_m of some common polymers

Polymer	T_m (K)	ΔH_m (kJ mol^{-1})	ΔS_m (J K^{-1} mol^{-1})
Polyethylene	414	8.0	19
Poly(tetrafluoroethylene)	672	5.7	8.5
Poly(ethylene oxide)	342	8.7	25
Polystyrene (isotactic)	513	9.0	18
Polystyrene (syndiotactic)	542	8.6	16
Poly(cis-1,4-isoprene)	299	4.4	15
Poly(trans-1,4-isoprene)	347	13	37

Fig. 3.8. Photograph comparing space-filling models of polyethylene (front) and poly-(tetrafluoroethylene) (PTFE).

in the melts. Suppose also that an attempt was made in advance to predict the conformation of this chain in the snapshot. Clearly the probability of this prediction being correct is much higher in the case of PTFE, when the chains are almost straight even in the melt, than in the case of polyethylene. This is in fact to say that the degree of disorder in and hence the entropy of molten polyethylene is much higher than those for molten PTFE. On the other hand crystallites of polyethylene and PTFE can be considered to have minimal disorder and hence roughly equal entropies. Thus the relatively low value of ΔS_m and hence the relatively high value of T_m for PTFE mainly reflects the relatively stiff nature of the chains in the molten state. This view will be extended in Section 9.3 to polymers with much higher values of T_m.

Poly(ethylene oxide) would be expected to have a more mobile chain even than polyethylene, as discussed in the preceding section, so that its larger value of ΔS_m and lower value of T_m in Table 3.3 are expected. Equally, the polystyrene backbone has bulky phenyl groups attached to every alternate carbon atom, which would be expected to stiffen it compared to polyethylene and thus give it the higher T_m.

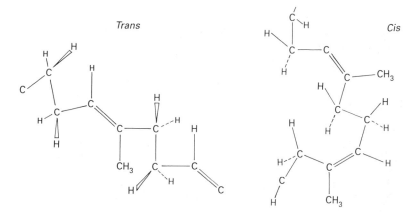

Fig. 3.9. Representations of the conformations of *trans*- and *cis*-poly(1,4-isoprene) chains which have all of the carbon atoms in the backbone in one plane.

The polyisoprenes in Table 3.3 are geometrical isomers, the *cis* configuration corresponding to natural rubber whilst the *trans* configuration is found in gutta-percha, another natural substance. Both have rather low values of T_m, not widely different, but the pairs of values of ΔH_m and ΔS_m are quite markedly different. Figure 3.9 represents the conformations of these chains which result in all of the backbone carbon atoms being in one plane. It is apparent that the poly(*trans*-1,4-isoprene) chain shown is directed along a single axis, running from top left to bottom right as drawn. Equally evidently the poly(*cis*-1,4-isoprene) cannot be linear whilst also keeping the backbone atoms in one plane. Thus the *trans* polymer is much more amenable to a crystalline phase, which is reflected in its much larger values of ΔH_m and ΔS_m than those of the *cis* polymer: these are consistent with a crystalline phase composed of close-packed aligned chains with relatively high interchain cohesion, which melts to give loose chains able to indulge in very free internal rotations. As mentioned before in connection with Table 3.2, the presence of C=C bonds in the backbone eases internal rotation about neighbouring C—C bonds in the melt, so that ΔS_m is rather large for the *trans* polyisoprene when melting involves the disappearance of a highly ordered crystalline phase. Poly(*cis*-1,4-isoprene) has twisting chains which are not amenable to packing together so that a rather disordered crystalline phase is all that can form, consistent with the low ΔH_m in Table 3.3. But the *fractional* changes in ΔH_m and in ΔS_m from *cis* to *trans* polymer are not very different from each other, so that they are expected to have T_m values which are quite close. Nevertheless the value of T_m of poly(*cis*-1,4-isoprene) is around ambient and crystallinity is

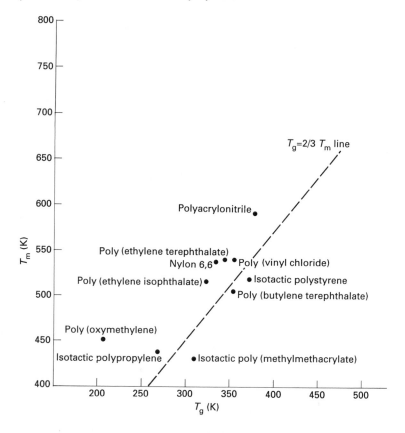

Fig. 3.10. Plot of T_m versus T_g with the points corresponding to several polymers and the line corresponding to T_g (K) = 2/3 T_m (K) shown.

usually insignificant in the corresponding elastomer (natural rubber). Gutta-percha is more like a thermoplastic, with significant crystallinity and consequently no useful elastomeric properties.

Both T_g and T_m increase as internal rotations in the bonds in the backbones of polymer chains become more hindered. In fact there is a useful approximate rule of thumb which states that T_g(K) is approximately equal to two-thirds (0.67) of T_m(K). Figure 3.10 indicates the extent to which this is borne out in the cases of several common polymers.

3.6 Measurement of T_g and T_m

The glass transition temperature (T_g) and melting temperature (T_m) can be determined most easily by differential thermal methods, specifically *differential thermal analysis (DTA)* and *differential scanning calorimetry (DSC)*.

The basis of DTA is that the polymer sample (typically 1–10 mg) in one of the matched aluminium cups has heat supplied at the same rate as an inert solid in the other cup, both being in nitrogen gas. The heating rate is controlled so that the temperature of the inert solid rises at a constant rate usually. The difference in temperature (ΔT) between the two cups is recorded using a thermocouple with one junction in each. When a transition takes place in the polymer, it will be reflected in an apparent change in the heat capacity, an increase if the transition is endothermic and a decrease if it is exothermic. Such a transition will be marked by a significant deviation in the otherwise smooth profile of ΔT versus time. The heat capacities of the polymer and inert solid will usually be different so that ΔT is not zero. DTA is a good technique for locating T_g (marked by a point of inflexion, as expected of a second order transition) and T_m (marked by a sharp trough as expected of a first order endothermic transition). The limitation of DTA is that it provides no quantitative information on the enthalpy change of melting (ΔH_m). That is where the DSC technique has its principal advantage which has resulted in its dominance over DTA for thermal investigations of polymer samples.

Differential scanning calorimeters have essentially the same basic components as differential thermal analysers. But in DSC systems the cups containing the polymer sample and the inert solid are provided with individual electrical heaters: the one heating the polymer sample supplies energy at the varying rates which are needed to keep ΔT equal to zero throughout whilst the other heater raises the temperature (T) of the inert solid at a linear rate. Thus, in DSC, it is the difference in electrical power (ΔQ) supplied to the two cups which is measured as a function of T and hence time (t). The *thermogram* which is displayed is a profile of the instantaneous rate of change of ΔQ (i.e. $d(\Delta Q)/dt$) against T. There is an important quantitative aspect of DSC thermograms: the integrated area between the profile and the baseline between the limits of T of a particular feature is directly related to the enthalpy-change for the corresponding transition in the polymer sample. Thus when the DSC system has been calibrated (say by using substances with known values of ΔH for say melting) enthalpy changes, such as ΔH_m for polymers, can be measured.

Figure 3.11 shows typical DSC profiles which have been obtained for samples of poly(ethylene terephthalate) (PET) treated in different ways

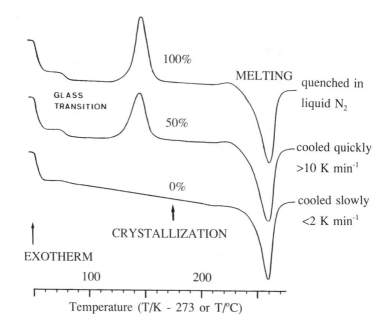

Fig. 3.11. Curves obtained by differential scanning calorimetry (DSC) on a 5.25 mg sample of poly(ethylene terephthalate) treated in different ways (right of curve) before the measurements. Reproduced by permission of Elsevier Science Publishers BV from Wiedemann, H. G., McKarns, T., and Bayer, G. (1990), Thermal characteristics of polymer fibres. *Thermochimica Acta*, **169**, 1–13.

beforehand. The top-most profile is for a totally amorphous sample obtained by rapid quenching of melted PET on dropping into liquid nitrogen at 77 K, which allows no time for the development of micro-crystallinity. The middle profile is for a PET sample which has been cooled at a moderate rate from the melt and hence has a substantial degree of crystallinity. The bottom-most profile is for a PET sample which has been cooled so slowly that it can be considered to have been annealed, i.e. its temperature has remained near to T_m long enough to develop around the maximum possible degree of crystallinity. The first phenomenon encountered with rising temperature is the glass transition, which appears as the points of inflexion on the upper two profiles but is not evident on the lowest one. Since this is a transition in the amorphous phase, it is perhaps not surprising that it almost fails to appear in the thermogram of the very highly crystalline sample. The next feature with increasing T is an exothermic peak, most developed on the top profile and not apparent in the bottom one. The underlying phenomenon is a form of crystallization, which is substantiated by observations of PET samples under a microscope. At around 425 K, PET chain segments in amorphous regions become sufficiently mobile that substantial parts of chains are able to extricate themselves from the 'spaghetti' and align with one another in a more cohesive crystalline manner, releasing heat as a consequence. Since the lowest profile corresponds to a sample with no scope for further crystallization, no exothermic peak appears at this temperature in this profile. Finally the features at highest temperatures correspond to endothermic melting of the crystallites. The upper edges of these indicate that T_m is about 540 K, whilst the integrated peak areas indicate that ΔH_m is around $22 \, \mathrm{kJ \, mol^{-1}}$, so that ΔS_m is approximately $41 \, \mathrm{J \, mol^{-1} \, K^{-1}}$.

3.7 Concluding remarks

This chapter may be concluded with a statement of general validity. For most polymers, the more regular is the chain at the molecular level, the higher will be the degree of crystallinity possible and the higher will be ΔH_m. This will not usually result in a bulk polymer with a high value of T_m, because features which increase the value of ΔH_m also tend to raise ΔS_m. Included in this is stereoregularity of chains with chiral centres in the backbone, a topic which will be discussed further in Section 8.1.

Exercises

3.1. Match each of the items (i)–(x) in the left-hand column with the most closely associated one from the right-hand column
      (i) elastomeric rubber   (a) natural rubber
     (ii) engineering PET   (b) above T_g only
    (iii) fibre draw ratios   (c) interchain chemical bonds
    (iv) DSC thermogram   (d) final stage of crystallite melting
     (v) nylon 6,6   (e) amorphous fibre-former
    (vi) polyacrylonitrile   (f) amorphous and crosslinked
   (vii) T_m (g) ΔH_m measurement
 (viii) segmental rotation   (h) usually four to six
    (ix) vulcanization   (i) chains of 150–300 CRUs usually
     (x) poly(cis-1,4-isoprene)   (j) interchain hydrogen bonds.

3.2. Consider a crystallite in polyethylene which is typically 10 nm across. On the basis that one polymer chain extends across this crystallite in its planar zigzag conformation, estimate the number of monomer residues of this chain which are in one traverse on the assumption that the chain geometry is the same as that given for polypropylene in Fig. 3.7.

3.3. Figure 3.12 represents the conformation which a form of poly(vinyl chloride) adopts in crystallites, where only the carbon and chlorine (larger) atoms are shown.
  (b) What is the tacticity of this chain?
  (b) How can the linear dimension given be explained?

0.51 nm   0.254 nm

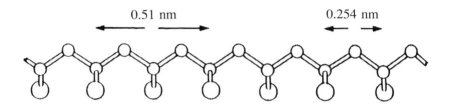

Fig. 3.12. Representation of the chain structure of the zigzag planar-backbone conformation of a poly(vinyl chloride). Reproduced by permission of the Royal Society of Chemistry from Bunn, C. (1975), Macromolecules—the X-ray contribution. *Chemistry in Britain*, **11**, 171–7.

3.4. Figure 3.13 represents the conformation adopted by the chain within crystallites in a form of polypropylene, with only the backbone carbon atoms and methyl groups (larger circles) shown.
  (b) What is the tacticity of the chain shown?
  (b) What does the linear dimension indicated correspond to?
  (c) Draw a view of this chain looking along the central (long) axis of the chain.

0.74 nm

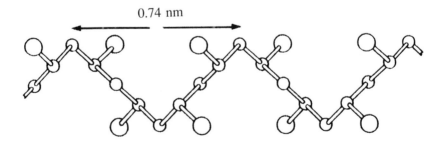

Fig. 3.13. Representation of the chain structure of the zigzag backbone conformation of a polypropylene. Reproduced by permission of the Royal Society of Chemistry from Bunn, C. (1975), Macromolecules—the X-ray contribution, *Chemistry in Britain*, **11**, 171–7.

3.5. Match each CRU structure in the left-hand column with the T_g value in the right-hand column which seems most appropriate

(i) $CH_2—C(CH_3)_2—CH_2—O—$ (a) 553 K

(ii)    (b) 295 K

(ii) $—CH_2—CH—$
   (c) 223 K

(iv)    (d) 424 K

(v)    (e) 353 K.

(Hint: compare these with related structures for which T_g values are given in the chapter and with one another.)

3.6. Match each CRU structure in the left-hand column with the most appropriate value of T_m from the right-hand column

(i) 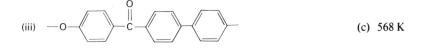   (a) 343 K

(ii) $—CH_2—CH(F)—$ (b) 713 K

(iii) 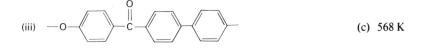   (c) 568 K

(iv) $—NH—(CH_2)_6—NH—C—(CH_2)_6—C—$ (d) 473 K

(v) $—CH_2—CH=C(Cl)—CH_2—$ (*cis*-form)   (e) 508 K.

3.7. The amount of PET concerned in Fig. 3.11 is 5.25 mg. Calibration of the area of the trough in the DSC profile corresponding to fusion showed that it corresponded to an amount of energy of 0.6038 J, expressed on the basis that the PET sample is fully crystalline.

(a) What is the value of ΔH_m for PET?

(b) How do the values of ΔH_m for polyethylene (Table 3.3) and PET compare when both are expressed in kJ kg^{-1}?

3.8. The heat of crystallization (ΔH_m) of PTFE is know to be directly proportional to $(\bar{M}_n)^x$, where x is a constant. Corresponding values are listed below

$\Delta H_m /(\text{kJ kg}^{-1})$	28.9	21.5	19.5	17.8
$\bar{M}_n$	9.85×10^5	4.53×10^6	7.50×10^6	1.20×10^7

(a) Evaluate x.

(b) Evaluate $\bar{M}_n$ for a sample with $\Delta H_m = 24.8$ kJ kg^{-1}.

(c) Comment on the significance of this relationship.

Source: Wliochowicz, A. and Sicigata, R. (1989). *British Polymer Journal*, **21**, 205–7.

4
Major techniques for analysis and structure determination

This chapter describes how various spectroscopic and scattering techniques may be used to reveal structural features of bulk polymers.

4.1 Infrared absorption spectroscopy

The infrared absorption spectra of many bulk polymers are almost unbelievably simple considering the large numbers of atoms incorporated into the molecular chains. Much complexity is precluded by the facts that large numbers of individual vibrations have virtually the same frequencies and that many vibrations are inactivated spectrally by selection rules.

Table 4.1 lists wavelengths (λ) and the equivalent wavenumbers ($\bar{\nu}$) which correspond to approximate spectral locations of the peaks of absorption bands of main interest in connection with some of the common polymers. Exemplifying polymers are also listed. In conventional infrared spectrometers, refracting prisms (typically of rock salt) or diffraction gratings are the central feature, acting by dispersing the infrared radiation spatially. The detector scans this so that radiation centred on a particular wavelength in a very small range impinges upon it at any instant. Figure 4.1 shows representative spectra; spectrum (a) is for pure poly(vinyl chloride) (PVC)

Table 4.1 Infrared absorption bands: locations, assignments, and examples

$\lambda(\mu m)$	$\nu(cm^{-1})$	Assignment	Exemplifying polymer
3.4	2940	C$-$H ⎫	
6.8	1470	C$-$H in CH$_2$ ⎬	Polyethylene
7.3	1370	C$-$H in CH$_3$ ⎭	
6.1	1640	>C=C<	Poly(*cis*-1,4-isoprene)
5.8	1720	>C=O ⎫	
8.9	1120	$-$C(=O)$-$O$-$ ⎬	Poly(methylmethacrylate)
6.2	1610 ⎫		
6.7	1490 ⎬	Aromatic ring	Polystyrene
13.3	750		
14.4	690 ⎭		
14.5	690	$-$C$-$Cl	Poly(vinyl chloride)
8.3	1210	C$-$F in CF$_2$	Poly(tetrafluoroethylene)

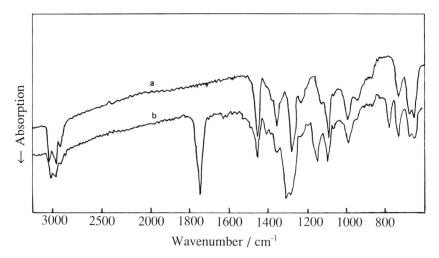

Fig. 4.1. The infrared absorption spectra of (a) pure poly(vinyl chloride) and (b) poly(vinyl chloride) plasticized with dioctylphthalate (di(2-ethylhexyl) phthalate). Reproduced with permission of the Division of Chemical Education, American Chemical Society from Chan, W. H. (1987). Plasticizers in PVC. *Journal of Chemical Education,* **64**, 897–8.

and the characteristic band listed in Table 4.1 is apparent near to the right-hand edge. The C—H band is also evident near to the left-hand limit. The lower spectrum is for a sample of PVC after an involatile liquid ester (a 'plasticizer', see Section 8.2) has been incorporated. The presence of this is most obvious in the appearance of the characteristic $>$C$=$O absorption at $1740\,\mathrm{cm}^{-1}$, similar to that of poly(methylmethacrylate) listed in Table 4.1.

Figure 4.2 provides another example of the information on the structure of a polymer which can be obtained from its infrared spectrum. This is for polypropylene which is predominantly syndiotactic. The solid arrows in the figure mark bands which are assigned specifically to the crystalline phase in the bulk. Also indicated by the dotted arrow is a weak feature which shows the presence of a small fraction of polypropylene residues which are joined in the unusual head-to-head configuration (Section 1.4).

Fourier transform infrared (FTIR) spectroscopy uses interferometry in the course of generating the spectra. In FTIR the detector monitors the entire spectrum at any moment rather than scanning it over a period of time as in conventional spectrometry. As a consequence, greater sensitivity and improved spectral resolution are obtained generally with FTIR instruments, which are also normally interfaced with a data processing facility which allows the generation of a difference spectrum between two actual spectra. Two instances of the results of this type of procedure appear in Fig. 4.3

Wavenumber / cm^{-1}

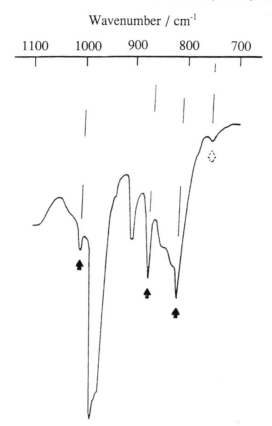

Fig. 4.2. Infrared absorption spectrum of syndiotactic polypropylene. Solid arrows indicate the bands ascribed to crystallinity, whilst the dotted arrow indicates the location of a band ascribed to the head-to-head linkage of propylene residues. Reproduced by permission of Hüthig & Wepf Verlag, Basel, form Kakugo, M., Miyatake, T., Naito, Y., Mizunuma, K., (1989). Microstructure of syndiotactic polypropylenes prepared with heterogeneous titanium-based Ziegler–Natta catalysts. *Makromolekulare Chemie*, **190**, 505–14.

for poly(methylmethacrylate)s. In both cases the difference spectrum represents the spectral consequences of the presence of crystallites in the samples. For example in the lower set of spectra, it is very clear that the absorption band centred on 1160 cm^{-1} is associated with the amorphous material only, since there is no obvious feature here in the difference spectrum assigned to the crystalline phase alone.

Raman spectroscopy can also provide information on the vibrations in solid polymers. The incident beam is usually visible light, often from a laser when it is typically the 514.5-nm-line emission from an argon ion laser with

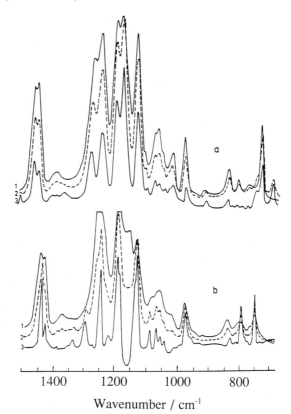

Wavenumber / cm^{-1}

Fig. 4.3. Infrared absorption spectra of films of (a) isotactic poly(methylmethacrylate)-α-CD$_3$ and (b) syndiotactic poly(methylmethacrylate)-α-CD$_3$, where the spectra labelled as 1 are for amorphous samples, 2 are partly crystalline samples and 3 are the spectra of the crystalline phase itself obtained in each case from spectrum 2 by digitally subtracting the corresponding spectrum of the amorphous phase. Reproduced by permission of Hüthig & Wepf Verlag, Basel from Schneider, B., Spěvaček, J., Straka, J., and Štokr, J. (1990). 'NMR and infrared study of crystalline forms of stereoregular poly(methylmethacrylates). *Makromolekulare Chemie, Macromolecular Symposia*, **34**, 213–26.

some 200 mW output power. In the scattering of this light by the polymer, energy equivalent to certain vibrational transitions is lost by some of the photons, so that the scattered light contains some higher wavelengths. Infrared absorption and Raman spectra are often complementary: the former responds best to bonds between different atoms, the latter to homopolar bonds such as C—C and C=C. As with the infrared absorption technique it is often simply a matter of finding a revealing band in the Raman spectrum which is a profile of the intensity of scattered light vs wavenumber difference between this and the incident light. For example in the Raman spectrum of polyethylene there is a narrow band centred on

Table 4.2 Densities (kg dm^{-3}) of solid phases of common polymers at 298 K

Phase (100%)	Polymer			
	Polyethylene	PET	PTFE	PEEK*
Crystalline	1.00	1.455	2.302	1.378
Amorphous	0.855	1.335	2.000	1.264

*Poly(etheretherketone)

1416 cm^{-1} which has been assigned to the bending vibrations of CH_2 units in the parts of the chains which are incorporated into crystallites. Another band centred on 1303 cm^{-1} is quite specific to the amorphous phase and is assigned to CH_2 twisting therein. The integrated intensities of these bands can be used to deduce the fractions of polyethylene samples which are in the different phases. This technique has been used for instances to show that 34 per cent is a representative degree of crystallinity in low density polyethylene (density = 0.92 kg dm^{-3}) whereas there can be up to 71 per cent crystallinity in high density polyethylene (density = 0.97 kg dm^{-3}). One important further feature to emerge is that there is a significant interfacial region between amorphous and crystalline phases with intermediate properties which makes up 15–20 per cent of most polyethylene samples.

Measurements of the densities of solid polymer samples by standard methods (for example pycnometry, gradient columns) can provide straightforwardly reasonably accurate estimates of their degrees of crystallinity. Values of the densities of pure amorphous and pure crystalline phases are needed and are readily available. Table 4.2 lists values for a few representative common polymers. Simple proportion is used to obtain the estimate of the degree of crystallinity of a bulk sample. This will however always be slightly in error on the basis of the neglect of any interfacial region between the two phases.

4.2 Nuclear magnetic resonance (NMR) spectroscopy

NMR spectroscopy has yielded truly spectacular insights into the structures of bulk solid polymers in recent years. Its very high resolution modes have allowed investigations of the sequences and stereoisomerism of monomer residues, as has already been indicated in Figs. 2.8 and 2.9, in which proton (^{1}H) NMR spectra of poly(methylmethacrylate)s were represented. There is no need here to do more in connection with the theory of NMR spectroscopy other than simply to accept that the chemical shifts (δ) are extremely sensitive to the number, type and spatial arrangement of the other atoms in the vicinity of that mainly responsible for a spectral feature.

^{13}C NMR spectroscopy has proved to be the most informative technique of all in recent years. The ^{13}C isotope has a natural abundance of 1.1 per cent only. But sensitivity is now high enough to allow signals from the few ^{13}C atoms amongst the dominant ^{12}C atoms to be easily detected. In comparison with ^{1}H NMR, the chemical shifts in ^{13}C NMR are considerably larger so that spectral features can in general be observed with much better resolution. Moreover the spectra are not complicated by ^{13}C—^{13}C coupling effects, simply because the sparsity of ^{13}C atom pairs in close enough proximity for these to be significant. Various other instrumental refinements, such as 'magic angle spinning' (MAS), 'cross polarization' (CP) and 'proton decoupling' (PD) are applied to sharpen up the otherwise broad features which appear in the spectra obtained with polymeric solids. Figure 4.4 illustrates the power of the technique. The polymer concerned is a block copolymer composed mainly of a chain of styrene monomer residues $((-CH_2-CH(Ph))_m-$, where Ph represents the aromatic ring) joined onto a chain of ethylene residues $(-CH_2-CH_2)_n-)$. The carbon atoms numbered 1 to 4 are the residues of the initiator (the isobutyl anion, $CH_3CH_2CH^-(CH_3)$, see Section 7.3).

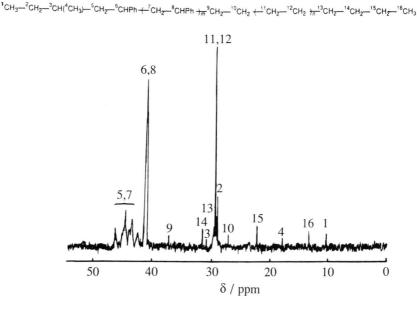

Fig. 4.4. ^{13}C NMR spectra of the polymer with the structure indicated above. Peaks are ascribed to the carbon atoms with the corresponding numbers appearing as superscripts before the C concerned in the structural formula. Reproduced by permission of the Royal Society of Chemistry from Endo, K. and Otsu, T. (1990). Novel synthesis of a block copolymer consisting of styrene and ethylene by means of active site modification. *Journal of the Chemical Society, Chemical Communications*, **19**, 1372–3.

The main point is that every type of carbon atom in the chain in terms of its immediate microenvironment has a different chemical shift and thus gives rise to a particular spectral feature.

In another instance ^{13}C NMR spectra have been obtained for the prepolymers formed during the initial stage of syntheses of urea $(CO(NH_2)_2)$–formaldehyde (HCHO) thermoset resins. Prepolymers are the partially polymerized molecules which do not possess the rigidity which follows the next stage in which they are linked into a giant network of the final polymer. The prepolymers are bound to have chemical structures which are the same as those of the final polymer, except that these have only developed to a limited extent. In the work concerned the prepolymers had been freeze dried and it was in this solidified form that the NMR spectra were taken. Two prepolymers were used: 'resin A' had been prepared using an acidic catalyst whereas an alkaline catalyst was used to produce 'resin B'. Figure 4.5 shows the ^{13}C NMR spectra (obtained with MAS and CP applied) of these prepolymers. The various features in these spectra are assigned in Table 4.3, which also gives data for the quantitative

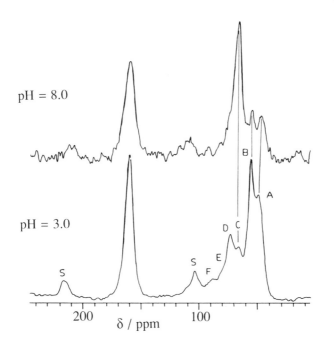

Fig. 4.5. ^{13}C NMR (with cross polarization and magic angle spinning) spectra of urea–formaldehyde prepolymers prepared at the pH values indicated. The assignments corresponding to the letters identifying peaks (S = spinning sideband) are given in Table 4.3. Reproduced from Jada, S. S. (1990). Solid-state ^{13}C-NMR studies of uncured urea–formaldehyde resins. *Journal of Macromolecular Science–Chemistry*, **A27** 361–75 by courtesy of Marcel Dekker, Inc.

Table 4.3 ^{13}C NMR data analysis of urea–formaldehyde prepolymers

Label	Assignment	Chemical shift (ppm)	Percentage of total carbon atoms	
			Resin A	**Resin B**
$>$N$-$C$-$N$<$ (carbonyl O)	Carbonyl carbon	160	41	33
A	$-$NH$-$CH$_2$$-NH-$	48.3	9.6	7.8
B	$-$C($=$O)$-$N$-$CH$_2$$-$ with CH$_2$, NH branch	54.5	39	7
C	HO$-$CH$_2$$-$N(H)$-$	65.3	5	50
D	HO$-$CH$_2$$-N<$	72.4	2	0
E	$-$CH$_2$$-O-CH_2$$-$	77.2	1	0
F	($-$CH$_2$$-O-CH_2$$-$O)$_n$$-$	88–90	2	2

Based upon data in Jada, S. S. (1990). Solid-state ^{13}C-NMR studies of uncured urea–formaldehyde resins. *Journal of Macromolecular Science–Chemistry*, **A27**, 361–75.

analyses furnished by deconvolution of the observed profiles into the various individual peaks. The large differences in the relative sizes of peaks B and C of the two prepolymers reveals that different main mechanisms are concerned in their formation. When resin A is formed under acidic conditions, the chains have a relatively high proportion of branching at nitrogen atoms on the basis of the structure assigned to B in Table 4.3. Also it seems likely that the major component species are terminated by primary amide groups ($-$C($=$O)$-$NH$_2$) rather than by methylol groups ($-$CH$_2$$-$OH). Resin B is indicated to have a much lower proportion of branched chains than does resin A, from the comparison of the fractions of the carbons atoms associated with peak B in the table. Methylol groups are apparently the major terminating groups in resin B.

The third example of the application of NMR methods to polymeric structure elucidation concerns a copolymer of vinylidene fluoride (CH$_2$$=CF_2$) and tetrafluoroethylene (CF$_2$$=CF_2$). The fluorine atom is sufficiently small that vinylidene fluoride units can be joined onto the chain in either way. Thus pentad sequences can be expected as follows: $-$CF$_2$CF$_2$CH$_2$CF$_2$CF$_2$$-$ (22022), $-$CH$_2$CF$_2$CH$_2$CF$_2$CF$_2$$-$ (02022), $-$CH$_2$CF$_2$CH$_2$CF$_2$CH$_2$$-$ (02020), $-$CF$_2$CH$_2$CH$_2$CF$_2$CF$_2$$-$ (20022), and $-$CF$_2$CH$_2$CH$_2$CF$_2$CH$_2$$-$ (20020), the codes in brackets using 2 for CF$_2$ and 0 for CH$_2$. Triad sequences on the same basis are: $-$CH$_2$CF$_2$CH$_2$$-$ (020), $-$CH$_2$CF$_2$CF$_2$$-$ (022), and CF$_2$CF$_2$CF$_2$$-$ (222). Figure 4.6 shows these features assigned to peaks in the proton (^{1}H) (a), ^{13}C (b), and ^{19}F (c) NMR spectra of the copolymer with molar composition 78 per cent

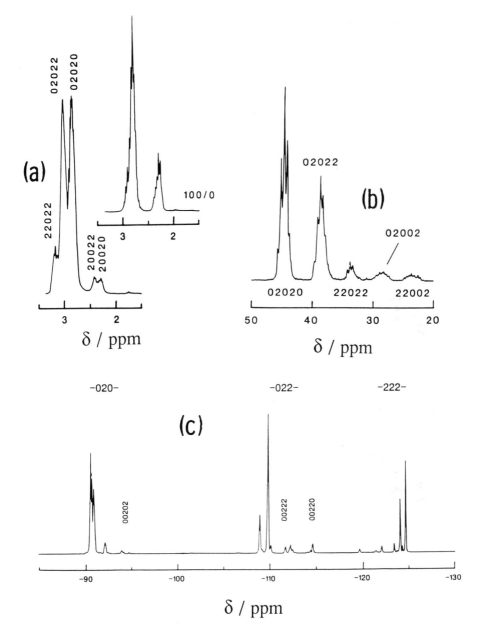

Fig. 4.6. NMR spectra of a copolymer of vinylidene fluoride and tetrafluoroethylene (78/22) in solutions. (a) ^{1}H (proton) spectra at 303 K using 5 per cent solutions in dioxane-d$_8$ (the inset is the spectrum of a pure poly(vinylidene fluoride) with an enhanced level (11.4 per cent) of head-to-head and tail-to-tail linkages (b) Proton-decoupled ^{13}C spectrum at 323 K using 20 per cent solution in acetone-d$_6$. (c) Proton-decoupled ^{19}F spectrum at 363 K using 7 per cent solution in dimethylformamide-d$_7$. Reproduced by permission of Elsevier Science Publishers BV from Cais, R. E. and Kometani, J. M. (1986). Structural studies of vinylidene–tetrafluoroethylene copolymers by nuclear magnetic resonance spectroscopy. *Analytica Chimica Acta*, **189**, 101–16.

$CH_2=CF_2$ and 22 per cent $CF_2=CF_2$ in solution in various solvents. Also shown as an inset in Fig. 4.6(a) in the ^1H spectrum of poly(vinylidene fluoride), showing a small peak assigned to head-to-head linkages ($-CH_2CF_2CF_2CH_2-$) and a large peak corresponding to the normal head-to-tail linkages (Section 1.4) ($-CH_2CF_2CH_2CF_2-$). The ^{19}F spectrum is apparently the best resolved which allows the quantitative estimation of the abundances of all of the pentads from the peak areas.

The examples used in this section are chosen to give indications of the power of the NMR techniques in general.

4.3 X-ray scattering

A considerable amount of information on the microscopic architectures of bulk polymers has come from the use of scattering techniques, including those based on X-rays, electrons, and neutrons as the scattered radiation or particles. X-ray scattering has proved to be the most important of these and two modes are discussed in this section.

X-rays are emitted from cathode-ray tubes in which a metal target is bombarded with electrons of very high energy. Copper is a typical target, when the X-rays are designated as copper $K\alpha$ radiation characterized by a wavelength of 0.15418 nm, of the same order as the lengths of chemical bonds. The scattered radiation is divided into two types in terms of the polymeric structural features which gave rise to its emergence at small ($<6^0$) or wide angles with respect to the incident beam.

4.3.1 Wide-angle X-ray scattering (WAXS)

The wide-angle scattering mode is commonly termed X-ray diffraction and it has given valuable information on the spatial arrangement of the atoms within crystalline materials. This scattering is referred to as coherent since it takes place without changes in wavelength or phase of the scattered radiation relative to the incident. Polymer crystals of sufficiently large dimensions to form the scattering target cannot be created normally. Many polymer samples are microcrystalline with random orientations of the crystallites, so that they scatter X-rays in much the same pattern as would a rotating macroscopic crystal when certain angles of scattering are preferred. A detector measuring the intensity of radiation would find peaks in the responses in these directions. But at the same time the amorphous polymer surrounding the crystallites would give rise to a continuum of scattered radiation. Figure 4.7 shows the X-ray diffraction profile versus the diffraction angle (2θ) obtained with a powdered sample of poly(vinyl chloride). The appearance is a smooth profile with superimposed broad peaks. The areas labelled as ϕ_c (three components) and ϕ_a are considered to be proportional to the weight fractions of crystalline and amorphous

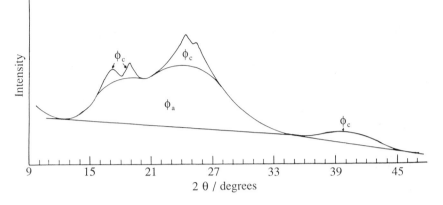

Fig. 4.7. Wide angle X-ray scattering (WAXS) profile of a poly(vinyl chloride) powder. Reproduced by permission of John Wiley & Sons Inc. from Obande, O. P. and Gilbert, M. (1989). Crystallinity changes during PVC processing. *Journal of Applied Polymer Science*, **37**, 1713–26. Copyright (1989) John Wiley & Sons Inc.

polymer respectively. On this basis, this PVC sample was around 9 per cent crystalline.

High degrees of crystallinity are produced when suitable polymers are drawn (Section 3.2). The crystallites have their polymer chains aligned closely with the length of the resultant fibre. WAXS performed with the fibre stretched across the X-ray beam would then be expected to produce a diffraction pattern characteristic of the crystalline phase. The unit cell is the minimum part of this which by its repetition allows the crystalline structure to be created. But it is only with some samples composed of very stereoregular or highly symmetrical polymer chains that the unit cell structure can be deduced from WAXS measurements alone.

In one instance a filament of poly(ethylene terephthalate) (PET) was drawn before being annealed under constant strain near to T_m, before drawing further, again at ambient temperature. Figure 4.8 shows the WAXS patterns produced on a photographic plate, after various treatments of the PET; λ is the draw ratio. Figure 4.9 shows models representing the dispositions of PET chains at the various stages.

There are evident changes in the WAXS patterns of Fig. 4.8 which are interpreted pictorially in Fig. 4.9. The starting filament is a predominantly amorphous material in which are embedded a large number of small crystallites containing many defects in their lattice structures. Thus Fig. 4.8(a) manifests a scattering pattern which results only from the directional effects of the PET filament; an intensity-measuring detector scanning from left to right would generate a smooth profile without significant peaks. The pattern of concentric rings (corresponding to peaks

(a) (b)

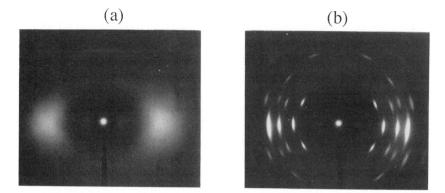

(c) (d)

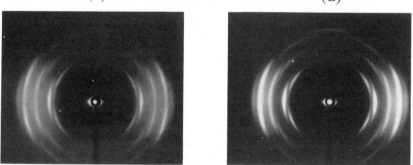

Fig. 4.8. Wide angle X-ray scattering (WAXS) patterns of a drawn poly(ethylene terephthalate) bristle, treated progressively as follows: (a) drawn at room temperature to draw ratio = 5, unannealed, (b) draw ratio = 5, after annealing at 533 K, (c) further drawing at room temperature to draw ratio = 20, no further annealing, (d) draw ratio = 20, after annealing at 533 K. Diameters of bristles: about 0.5 mm for (a) and (b), about 0.3 mm for (c) and (d). Reproduced from Fakirov, S. and Evstatiev, M. (1990). New routes to polyethylene terephthalate with improved mechanical properties. *Polymer*, **31**, 431–4. by permission of the publishers, Butterworth–Heinemann Ltd (Copyright (1990).

being superimposed on the left-to-right scan of intensity) develops with annealing and further drawing (Fig. 4.8(b) and (c)): this is indicated in Fig. 4.9(b) and (c) to result from the increasing alignment of chains, i.e. the rising degree of crystallinity.

It is clear that annealing near to T_m (typically 265°C (538 K)) induces considerable development of crystallinity. Part of this action is the further progress of step-growth polymerization in zones where additional

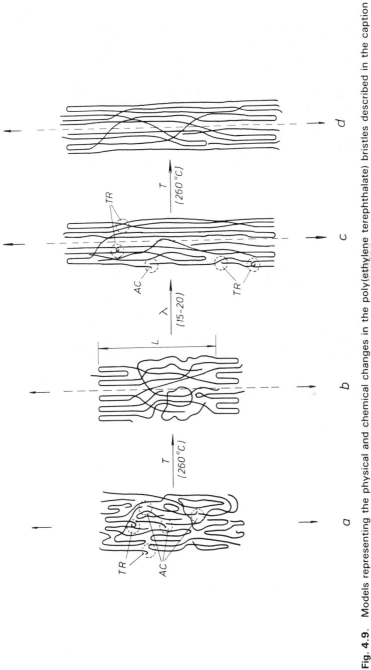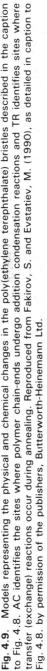

Fig. 4.9. Models representing the physical and chemical changes in the poly(ethylene terephthalate) bristles described in the caption to Fig. 4.8. AC identifies the sites where polymer chain-ends undergo addition condensation reactions and TR identifies sites where transfer (exchange) reactions occur during annealing. Reproduced from Fakirov, S. and Evstatiev, M. (1990), as detailed in caption to Fig. 4.8. by permission of the publishers, Butterworth-Heinemann Ltd.

condensations (AC) or transesterification reactions. (TR) (see Section 5.3.2) have taken place from (a) to (b) and also from (c) to (d) as indicated in Fig. 4.9. AC and TR actions result in the elimination of defects in crystallites associated with such microscopic features as chain entanglements or chain ends. This example will be discussed further in the next subsection.

The examples above illustrate the general level of information which is gained by WAXS measurements. In most instances the main quantitative result expected is simply the degree of crystallinity of a sample. But on other occasions the qualitative diagnosis of the presence or absence of crystallinity will be useful in itself.

4.3.2 Small-angle X-ray scattering (SAXS)

The main characteristic of the X-ray intensity scattered at small angles is diffuseness. But the intensity versus angle profile may show some structure

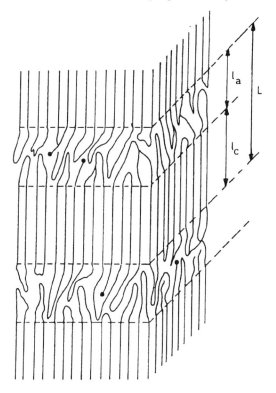

Fig. 4.10. Representation of the typical lamellar structure of partially-crystalline poly(ethylene terephthalate), in which L is termed the long spacing, l_c is the thickness of the crystalline layer, and l_a is the thickness of the amorphous layer. Reproduced with the permission of Comité van Beheer van het Bulletin v.z.w. from Huyskens, P., Groeninckx, P., and Vandevyvere, P. (1990). What rules the melting point of semi-crystalline polymers? *Bulletin des Sociétés Chimique Belges*, **99**, 1011–17.

which can be used to deduce such features as dimensions of crystallites or other aggregates, which are the scattering entities, and the modes of chain packing in the crystalline regions. At this point it is appropriate to discuss the types of crystallinity which can occur.

Many of the most important thermoplastics, including polyethylene, polypropylene and PET, have a characteristic folding of chains which is associated with the microcrystalline regions. It gives rise to a lamellar-like crystallite, typical lamellae being a few micrometres broad (i.e. in the directions perpendicular to the aligned chain axes) but with limited thickness, 0.01–0.3 μm covering most instances. Figure 4.10 represents this type of structure within a fibre, showing lamellae, in which the aligned chains are represented simply as parallel lines, with intervening amorphous regions above and below. The various thickness parameters are defined in the diagram. It is the long spacing (L) which may be deduced using SAXS.

Figure 4.11 shows the SAXS profile of intensity vs distance (h) from the central spot for the PET fibre concerned in Fig. 4.8. What is obvious is that only profile (b) shows a distinct maximum. The absence of maxima in profiles (a), (c), and (d) is interpreted as indicating the absence in the fibre

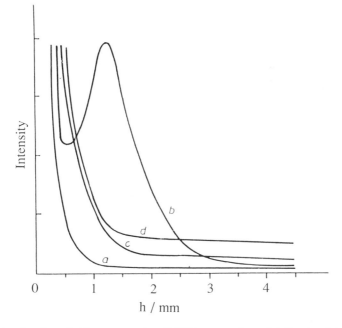

Fig. 4.11. Small angle X-ray scattering (SAXS) curves for a bristle of poly(ethylene terephthalate) subjected to treatments (a)–(d) described in the caption to Fig. 4.8. Reproduced from Fakirov, S. and Evstatiev, M. (1990), as detailed in the caption to Fig. 4.8. by permission of the publishers, Butterworth-Heinemann Ltd.

at these stages of regular patterns of alternating regions which differ substantially in density (Table 4.2). Conversely the annealing of the fibre after moderate drawing is indicated to induce disentanglement of chains and subsequent crystallization in lamellae separated by amorphous regions as is apparent in Fig. 4.9(b), even if the process is incomplete. Other evidence of the thermal induction of such crystallization processes in PET is manifested in Fig. 3.11. The relatively high degree of crystallinity achieved at this stage is apparent in Fig. 4.8(b), which approaches the pattern composed simply of spots which would be expected if a fully crystalline solid was the target. The PET fibre concerned in parts (d) of these diagrams is likely to have a higher degree of crystallinity than does that in parts (b), but the absence of a maximum in profile (d) in Fig. 4.11 confirms that the arrangement of chains does not create lamellae in this very stretched fibre. This accords with the view represented in Fig. 4.9(d) that the chains are highly orientated along the length of the fibre with individual chains having strong cohesion to their neighbours by virtue of the many aligned segments. The scattered intensity depends upon the density of the sample so that its marked increase with profile (d) in Fig. 4.11 reflects the maximum density which is achieved in the most crystalline sample of the PET.

A second instance of the applications of SAXS techniques concerns a block copolymer, referred to as SEBS, in which the central portion of each chain is poly(butadiene) and the ends are poly(styrene). Samples of SEBS were lightly sulphonated, so that anionic sulphonate groups ($-SO_3^-$) are attached to a minor proportion (2.5–8 per cent) of the styrene residues of each chain. The associated cations could be Na^+, when the polymer is designated as xNa$-$SBC, where x is the mole fraction in percentage terms which have a sulphonate group attached and SBS is an acronym for sulphonated block copolymer. When Zn^{2+} was the counterion, xZn$-$SBC was the designation. Figure 4.12 shows SAXS profiles of representative samples of three materials, in terms of the wavevector (q) which is evaluated by $4\pi \sin \theta / \lambda$, where λ is the X-ray wavelength (0.143 nm in this case). The corresponding scattering angles (2θ) range from 0.13^0 to 7.8^0 as q goes from 0.1 to 6.0 nm^{-1}. In Fig. 4.12(a), all three forms show a maximum in the profiles at very small angles, the region in which SAXS responds to the largest microstructural features, indicative of entities 20–30 nm across. This is what is expected of SEBS because it is a thermoplastic elastomer (see Section 8.4) in which the polystyrene chain sections separate out to form spherical crystallites embedded in an amorphous phase of poly(butadiene). This feature is not affected by sulphonation. Figure 4.12(b) reveals that the SBC forms differ from SEBS at smaller dimensions (i.e. larger values of q). In fact, these SBC polymers are ionomers (see Section 8.4) in which the charge centres separate out to form domains, typically 3–4 nm across, of an ion-rich microphase. It is the different X-ray scattering ability of the domains from the polystyrene

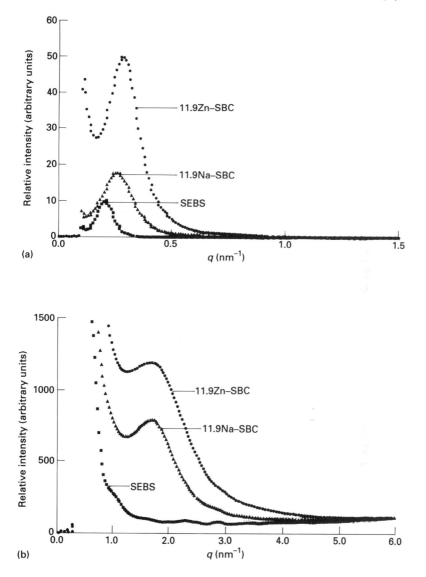

Fig. 4.12. Small angle X-ray scattering (SAXS) curves for a styrene-(ethylene–butene-copolymer)-styrene (SEBS) triblock copolymer which is lightly sulphonated (11.9 mol per cent) and subjected to exchange of the acidic hydrogen atoms for metal (M) cations, to give forms (ionomers) designated as 11.9M-SBC (sulphonated block copolymer) with M = Na and Zn. Reproduced from Weiss, R. A., Sen, A., Pottick, L. A., and Willis, C. L. (1990). Block copolymer ionomers: thermoplastic elastomers possessing two distinct physical networks. *Polymer Communications*, **31** 220–3. by permission of the publishers, Butterworth-Heinemann Ltd.

crystallites in which they are embedded which gives rise to the main peaks shown in Fig. 4.12(b). The main point for the present is the power of SAXS to reveal the differences in microstructure. Electron microscope photographs presented later (Fig. 9.10) show such domains more directly in similar ionomers.

One other instance in which WAXS and SAXS have been used to provide complementary information deserves to be mentioned. The material concerned was a stretched fibre of poly(para-phenylene terephthalamide), more familiar perhaps via its trade name of Kevlar (see Section 9.2). WAXS investigation of typical Kevlar fibres gave scattering patterns consistent with high degrees of crystallinity (70–80 per cent). But SAXS investigations of the same fibres revealed no evidence of a long spacing (L) like that in PET fibres (Fig. 4.10); there was no maximum in the profile like that in Fig. 4.11(b), even when the drawn fibres had been annealed thoroughly. This SAXS result indicates the complete absence of chain folding in Kevlar, which is expected when its chains are rigid and rod-like. On the other hand the WAXS result reflects the substantial alignment of these chains along the fibre length which promotes crystallinity in Kevlar.

The aim of this section has been to show how X-ray scattering techniques may be applied to give important information about microstructures of polymers, even when a complete structural analysis is not possible.

Exercises

4.1. Urea-formaldehyde prepolymers have been prepared in media with pH values of 7.5 and 3.2. Bands centred on wavenumbers of 3015, 2960, and 2900 cm^{-1} have been assigned to C—H vibrations in $-CH_2-O-CH_2-$, $-CH_2-N-CH_2-$, and $-CH_2OH$ groups respectively. In conjunction with the corresponding NMR results (Table 4.3), interpret the relative intensities in each band for the prepolymers synthesized at pH $= 3.2$ and 7.5 given below

Wavenumber (cm^{-1})	3015	2960	2900
pH $= 3.2$	1.4	38.5	5.5
pH $= 7.5$	2.3	6.9	50.0

Source: Jada, S. S. (1988). *Journal of Applied Polymer Science*, **35**, 1573–92.

4.2. Copolymers of ethylene and 1-hexene can be analysed by infrared absorption spectroscopy to establish their composition in terms of monomer residues using characteristic absorptions at wavenumbers of 1380 and 722 cm^{-1} respectively. The following corresponding values of absorbance (A) ratios and hexene residue content (Hx) were measured

A_{1380}/A_{722}	0.078	0.152	0.255	0.366	0.488
Hx/(mol%)	2.00	4.00	6.00	8.00	10.00

What is the percentage of hexene residues in the copolymers which yield the absorbance ratios $A_{1380}/A_{722} = 0.063$, 0.210, and 0.465 respectively?

Source: Nowlin, T. E., Kissin, Y. V., and Wagner, K. P. (1988). *Journal of Polymer Science, Part A*, **26**, 755–64.

4.3. Poly(vinyl alcohol) (PVA) $(-CH_2-CH(OH))_n$ has been prepared with an average degree of polymerization of 12 800. Figure 4.13 shows ^1H NMR spectra measured with the PVA (a) in solution in fully deuterated dimethyl-sulphoxide (DMSO-d_6) and (b) in solution in DMSO-d_6 when trifluoroacetic acid ($CF_3C(=O)OH$) has been used to convert all $-OH$ groups to trifluoro-acetate groups in monomer residues which are linked head-to-tail, leaving those on head-to-head linkages unreacted for steric reasons. Estimate (using

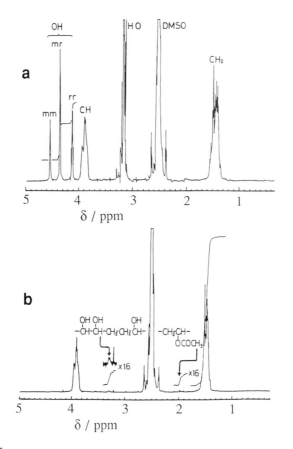

Fig. 4.13. ^1H (proton) nuclear magnetic resonance spectra of poly(vinyl alcohol) dissolved in fully deuterated dimethyl sulphoxide (a) itself and (b) with added trifluoroacetic acid. The groups and structures assigned to the various peaks are indicated. Integrated profiles amplified by a factor of 16 are shown within (b). Reproduced with permission of the Society of Polymer Science, Japan from Yamamoto, T., Seki, S., Fukae, R., Sangen, O., and Kamachi, M. (1990). High molecular weight poly(vinyl alcohol) through photo-emulsion polymerizations of vinyl acetate. *Polymer Journal*, **22**, 567–71.

the assignments in the diagram and measurements with a ruler as appropriate)

(a) the value of $\bar{M}_n$ of this PVA sample,

(b) the approximate proportions of isotactic, syndiotactic, and atactic triads,

(c) the proportion of head-to-head linkages of monomer residues.

Source: as in caption of Fig. 4.13.

4.4. The numbers of methylene ($-CH_2-$) and methyl ($-CH_3$) groups in polyethylene are directly proportional to absorbances (A) at wavenumbers of 2955 cm^{-1} and 2928 cm^{-1} respectively. The fraction (x) of carbon atoms which are in methyl groups can be measured by ^{13}C NMR spectroscopy. Measurements have been made on various samples of a commercial form of polyethylene, known as LLDPE (see Section 7.4), which is a copolymer of ethylene and 1-octene and is a linear chain with short branches of uniform length located on residues of the latter monomer. Corresponding values of the ratio of absorbances and x are listed below

A_{2955}/A_{2928}	0.0462	0.0639	0.100
x	2.035	2.815	4.410

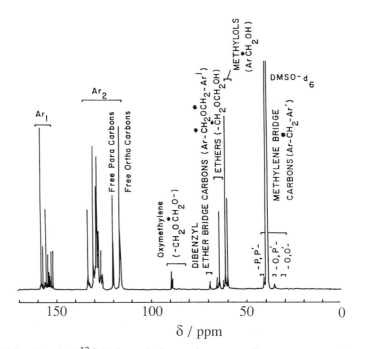

Fig. 4.14. A typical ^{13}C high resolution nuclear magnetic resonance spectrum of a resole. Ar$_1$ locates the peaks assigned to the carbon atoms which are in aromatic rings and are attached directly to OH groups, and Ar$_2$ locates peaks assigned to the other carbon atoms in aromatic rings. In other peak labelling, the particular carbon atom in the structural unit which gives rise to the peak is marked with an asterisk. Reprinted by permission of John Wiley & Sons, Inc. from So, S. and Rudin, A. (1990). Analysis of the formation and curing reactions of resole phenolics. *Journal of Applied Polymer Science*, **41**, 205–32. Copyright (1990) John Wiley.

Use these data to evaluate the average number of branches per chain in a sample of this LLDPE with $\bar{M}_n = 9.0 \times 10^4$ for which $A_{2955}/A_{2988} = 0.0227$.

Source: Housaki, T., Satoh, K., Nishikida, K. and Morimoto, M. (1988). *Makromolekulare Chemie, Rapid Communications*, **9**, 525–8.

4.5. Figure 4.14 shows the ^{13}C NMR spectrum of a resole type of phenol-formaldehyde resin which was obtained in the presence of DMSO-d$_6$ which served as an internal standard for chemical shifts (δ). Assignments are given above each feature, with an asterisk marking the specific carbon atom concerned. Ar$_1$ denotes the aromatic ring carbon atom which is attached directly to the phenolic —OH group whilst Ar$_2$ denotes any of the other carbon atoms in the aromatic ring. What may be deduced concerning the chemical structure of this resole?

Source: as in caption to Fig. 4.14.

4.6. Figure 4.15 shows X-ray diffractograms of films of various polyimides, denoted by (P), (B), and (O). What may be deduced?

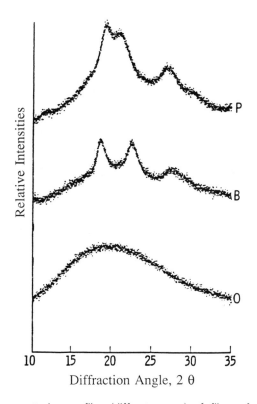

Fig. 4.15. X-ray scattering profiles (diffractograms) of films of three polyimides. Reprinted by permission of John Wiley & Sons, Inc., from Hergenrother, P. M., Wakelyn, N. T., and Havens, S. J. (1987). Polyimides containing carbonyl and ether connecting groups. *Journal of Polymer Science, Part A: Polymer Chemistry*, **25**, 1093–1103. Copyright (1987) John Wiley & Sons Inc.).

4.7.   Given that the densities of crystalline, interphase, and amorphous regions in polyethylene are 1,000, 0.964, and 0.855 kg dm^{-3} respectively (a) calculate the density of a sample in which Raman spectroscopy has been used to measure fractions of 0.42 and 0.39 of crystalline and amorphous phases respectively, (b) what percentage error would be made in assessing the crystalline fraction from the density calculated in (a) if the interfacial phase was assumed to be insignificant?

   Source: Földes, E., Keresztury, G., Iring, M., and Tüdős, F. (1991). *Die Angewandte Makromolekulare Chemie*, **187**, 87–99 (in English).

4.8.   Figure 4.16 shows WAXS profiles of poly(ethylene terephthalate) powder (a) as obtained from fibres in a fabric and (b) after melting at 548 K and then quenching in water at 273 K. Compare these with Fig. 4.8 and give a general interpretation.

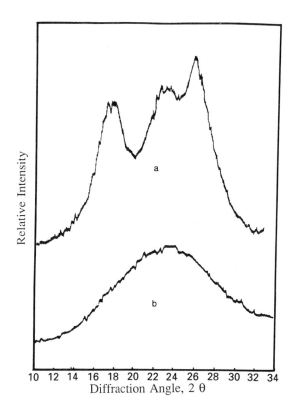

Fig. 4.16.   Wide-angle X-ray scattering (WAXS) profiles of poly(ethylene terephthalate) (a) in the form of ground spun-fibre and (b) in the form of an amorphous sample prepared by melting the fibres at 548 K and rapid quenching in water at 273 K. Reprinted by permission of John Wiley & Sons, Inc., from Hsieh, Y-L. and Mo, Z. (1987). Crystalline structures of poly(ethylene terephthalate) fibers. *Journal of Applied Polymer Science*, **33**, 1479–85. Copyright (1987) John Wiley & Sons, Inc.).

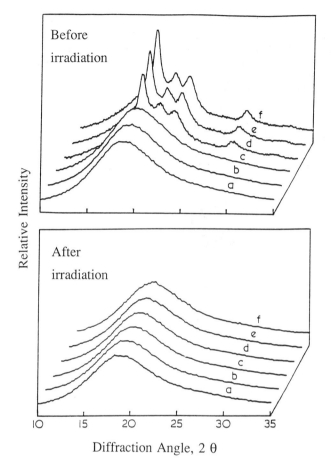

Fig. 4.17. Wide angle X-ray scattering (WAXS) profiles of a sample of poly(etherether ketone) (PEEK) measured at temperatures of 293 K(a), 373 K(b), 423 K(c), 473 K(d), 523 K(e), and 573 K(f) before irradiation (upper set of profiles) and after irradiation totalling 5000 Mrad using a 2 MeV electron beam in air at room temperature. Reproduced from Yoda, O. (1984). The radiation effect on non-crystalline poly(aryl-ether-ketone) as revealed by X-ray diffraction and thermal analysis. *Polymer Communications*, **25**, 238–40.

4.9. Figure 4.17 shows a series of X-ray diffratograms of initially amorphous poly(etheretherketone) (PEEK) before (upper diagram) and after (lower diagram) irradiation with high-energy electrons (5000 Mrad). The samples were heated progressively and the profiles shown correspond to highest temperatures reached of (a) 293 K, (b) 373 K, (c) 423 K, (d) 473 K, (e) 523 K, and (f) 573 K. What are the effects of (i) heating the unirradiated sample and (ii) irradiation of the sample and how many these be interpreted?

4.10. The characteristic peak in the SAXS profile for an ionomer is centred on values of the wavevector (q) ranging from 1.5 to 2.5 nm^{-1} for different ionomers. What are the corresponding spacings (d) between ion-rich domains if it is assumed that Bragg's Law applies to these peak locations in the form $\lambda = 2d \sin \theta$. The incident radiation is copper Kα with $\lambda = 0.154_{18}$nm.

Source: Yarusso, D. J. and Cooper, S. L. (1983). *Macromolecules*, **16**, 1871–80.

5
Step-growth polymerizations

Some introductory remarks on this topic have been made in Section 1.4, with examples of step-growth polymerizations shown in Fig. 1.8. This chapter discusses these types of processes in more detail.

5.1 General features

In step-growth polymerization, a single elementary reaction act, such as the creation of an ester linkage with elimination of a water molecule (i.e. a condensation) is repeated often to eventually build up the long chains. *Linear* chains are synthesized when the monomer molecules are *bifunctional* (i.e. have only two reactive groups) whereas *network* polymers are usually formed when at least one of the monomers has a higher functionality than two. Even when the reaction has proceeded as far as is possible, the chains or network structures have unreacted functional groups at the ends or edges.

For simplicity, at first consider a single monomer which is bifunctional with different and mutually reactive groups at the ends of the molecule, such as $HO----COOH$ where $----$ represents a chain of $-CH_2-$ groups typically. This monomer will undergo self-esterification, which is the first act of its step-growth polymerization represented as

$$HO----COOH + HO----COOH \rightarrow$$
$$HO----C(=O)-O----COOH + H_2O.$$

In more general notation, this type of monomer could be denoted as AB in terms of its functionalities only. Two joined monomer residues could be denoted as ABAB (or $(AB)_2$), when the above reaction would be written as

$$AB + AB \rightarrow ABAB \text{ (or } (AB)_2).$$

The eliminated small molecule (water in the case above) is now ignored for simplicity. The sequence of steps which results in the growth of a linear chain can be represented as

$$
\begin{array}{lllll}
AB & + & (AB)_2 & \rightarrow & ABABAB \text{ (or } (AB)_3) \\
(AB)_2 & + & (AB)_2 & \rightarrow & ABABABAB \text{ (or } (AB)_4) \\
AB & + & (AB)_3 & \rightarrow & (AB)_4
\end{array}
$$

$$(AB)_2 \;\; + \;\; (AB)_3 \;\; \rightarrow \;\; ABABABABAB \;\; (\text{or } (AB)_5)$$
$$\ldots \;\; + \;\; \ldots \;\; \rightarrow \;\; \ldots$$
$$(AB)_r \;\; + \;\; (AB)_s \;\; \rightarrow \;\; (AB)_{r+s}$$

$(AB)_r$ and $(AB)_s$ represent the molecular chains composed of r and s monomer residues respectively. In this case the monomer residue and the constitutional repeating unit (CRU) are the same. The rates of each of the steps represented above are proportional to the products of the concentrations of the reactants, for example $[(AB)_r][(AB)_s]$ in the last (general) step. When the system commences with only AB present, it is going to be a considerable time before concentrations of molecules with relatively long chains build up to significant levels. It is therefore clear that significant yields of $(AB)_{r+s}$ with large values of $(r + s)$ can only appear in the later stages of step-growth polymerizations. There are effectively no long chains present in the early stages.

Consider the situation when step-growth polymerization with a single monomer with two different and mutually reactive functional groups has proceeded until a fraction p of all of these groups (say $-OH$ and $-COOH$) has reacted. For each pair of these groups eliminated, two separate molecules have become one and thus p also corresponds to the fractional reduction in the number of molecules. Thus if N_0 was the number of monomer molecules added to the system and N is the number of separate molecules (of any size) later, the corresponding p is defined as $(N_0 - N)/N_0$ (when N_0 and N are very large numbers) leading to the equation

$$N = N_0(1 - p). \tag{5.1}$$

Since N_0/N is the original number of molecules divided by the remaining number, it is equal to the average number of monomer residues per molecule, i.e. the average degree of polymerization denoted by $\overline{DP}$. Hence eqn (5.1) leads to

$$\overline{DP} = N_0/N = 1/(1 - p). \tag{5.2}$$

This is known as the Carothers equation. It confirms immediately the fundamental point that a large extent of reaction (i.e. p values approaching 1.00) is essential for the existence of significant numbers of long chains in a synthesized step-growth polymer. Table 5.1 illustrates this with corresponding values of $\overline{DP}$ and p. Only the highest value of $\overline{DP}$ shown in the table corresponds to the average chain being a true polymer (Section 3.1).

The distribution of molecular masses about the average is also of interest. The analysis involves an approach in which it is necessary to recognize the equivalence of the probability of a defined result occurring in one random

Table 5.1 Corresponding values of $\overline{DP}$ and p via eqn (5.2)

p	0.5	0.75	0.90	0.95	0.99
$\overline{DP}$	2	3	10	20	100

event and the fraction of times that it will be achieved in a large number of such events. Consider rolling a dice, when each of the six faces is equally likely to be on top when it comes to rest. Thus the probability of this being say the 'one' is 1/6 in one throw. Equally if a very large number of throws of a single dice was made, it would be predicted that one sixth of these would show the 'one' face. Alternatively if many identical die were thrown simultaneously, one-sixth of them would be expected to stop with the 'one' face on top.

Imagine that there is a microscopic-scale device which has the ability to discriminate between the two functional groups when it is put into the step-growth polymerization system with monomer AB. Suppose that this device has the ability to attach itself at random to an unreacted functional group, say A, when the extent of polymerization is p. Let A^* identify this particular group. The value of p measures the fraction of the functional groups present at the start which have reacted; it also then expresses the probability that the B group of A^*B, when A^* has been selected at random, will have reacted. This means that p measures the probability that A^* is part of a molecule which is *at least* two monomer residues long, i.e. that the structure at the end of the molecule selected is A^*BAB. Now what is the probability that the B group at the right here has also reacted, i.e. that the end structure is A^*BABAB and is *at least* three monomer residues long? Probabilities combine by multiplication; this means that the probability that two results will occur simultaneously is the product of the individual probabilities for each. In the case of two dice, the probability of throwing two 'ones' is $(1/6)^2 = 1/36$. Thus the probability the chain terminated by A^* is at least three units long is the product of the probability (p) that A^* is part of A^*BAB multiplied by the probability (also p) that the B group at the right has reacted, i.e. $p \times p = p^2$. The general expression becomes apparent. The probability that the end structure is $A^*BABABAB$, i.e. that the selected molecule is *at least* four monomer residues long is evidently $p \times p \times p = p^3$. Thus the probability that the molecule terminated by A^* is composed of *at least* x monomer residues is $p^{(x-1)}$ by logical progression. Then what is the probability that this molecule is *exactly* x monomer residues long? Again this is specified by the product of the individual probabilities of two necessary conditions. The first is evidently the above probability that the chain is *at least* x monomer residues long. The second is the probability that the B group at the right of the

A*BAB————ABAB chain composed of x AB units has *not* reacted, which is $1 - p$ when p is the probability that it has reacted. The combination is $(1 - p)p^{(x-1)}$ which measures the probability that a molecule selected at random is exactly x monomer residues long. In view of the numerical equality of the probability of this chain length being obtained in the selection of one molecule to the fraction of molecules amongst the large number present which have this chain length, the fraction of *all* molecules which are $(AB)_x$ in a step-growth polymerization system is also $(1 - p)p^{(x-1)}$. So when N_x is defined as the number of $(AB)_x$ molecules and N is the number of all molecules (ranging from AB to $(AB)_\infty$) in the synthesized system, the equation for the number fraction is

$$N_x/N = (1 - p)p^{(x-1)}. \tag{5.3}$$

When eqn (5.2) is used to substitute N_0 for N, the result is

$$N_x/N_0 = (1 - p)(1 - p)p^{(x-1)}$$

from which is obtained

$$N_x = N_0(1 - p)^2 p^{(x-1)}. \tag{5.4}$$

Next an expression for the fraction of the total weight of a step-growth polymer which is composed of chains exactly x monomer residues long is

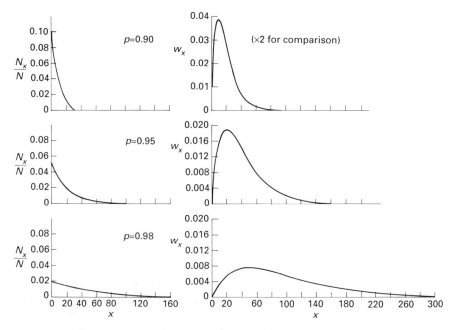

Fig. 5.1. Theoretical profiles of the number fraction (N_x/N) *(left)* and weight fraction *(W_x)* versus the number of monomer residues (x) in the polymer chains for extents of polymerization (p) in the step-growth polymerization of one bifunctional monomer.

required. The relative weight of $(AB)_x$ molecules is xN_xM, where M is the relative molecular mass of the monomer residue. The total mass of all monomer residues in the system is effectively equal to N_0M. Then the weight fraction (w_x) of the chains which are exactly x monomer residues long is expressed by the equation

$$w_x = xN_xM/N_0M = xN_x/N_0 = x(1 - p)^2p^{(x-1)} \qquad (5.5)$$

when eqn (5.4) is used to substitute N_x.

Figure 5.1 shows representative profiles of the number fractions (a) and weight fractions (b) versus x for three values of p, the extent of polymerization.

It is evidently essential to exceed $p = 0.95$ when only molecules with x exceeding 50 are true polymer chains. The number fraction profiles on the left reveal that monomer is the most abundant individual species, even when p is 0.98. The fraction of molecules which are truly polymeric is insignificant for $p = 0.90$, very small for $p = 0.95$ and still only around 0.37 when the reaction is 98 per cent completed. It is the weight fraction however which is the more important parameter for the production of these polymers on an industrial scale. The 'polymer weight fraction', denoted by Σw_{50+}, might be defined as the fraction of the yield by weight which has $x \geq 50$ and is equal to the fraction of the integrated area under the w_x profiles which lies to the right of $x = 50$. Table 5.2 lists values of Σw_{50+} for the profiles on the right of Fig. 5.1. This establishes that the product of step-growth reaction can only be termed polymeric in reality when p moves very close to 1.000.

The polydispersity index ($\bar{M}_w/\bar{M}_n$) of the total yield of a step-growth polymerization usually approaches 2.000. To explain this, $\bar{M}_n$ and $\bar{M}_w$ need to be expressed in terms of M and p. $\bar{M}_n$ is simply the product of the average number of monomer residues ($\overline{DP}$) and the relative molecular mass (M) of each. Hence using eqn (5.2) the resultant equation is

$$\bar{M}_n = \overline{DP} \cdot M = M/(1 - p). \qquad (5.6)$$

$\bar{M}_w$ is equated to $\Sigma w_x xM$ and with substitution for w_x using eqn (5.5), an equation is produced

$$\bar{M}_w = M(1 - p)^2\Sigma(x^2p^{(x-1)})$$

where the sum is to be taken over all values of x. The sum to infinity of

Table 5.2 The variation of polymer weight fractions (Σw_{50+}) with the extent of polymerization (p)

p	0.90	0.95	0.98
Σw_{50+} (%)	3.7	28.5	72.0

the series $x^2 p^{(x-1)}$ is equal to $(1 + p)/(1 - p)^3$, so that with incorporation of this the resultant equation is

$$\bar{M}_w = M(1 + p)/(1 - p). \tag{5.7}$$

Thus from eqns (5.6) and (5.7) the polydispersity index is given by

$$\bar{M}_w/\bar{M}_n = 1 + p \tag{5.8}$$

and its value approaches 2.000 as p tends towards 1.000.

Finally in this section it needs to be emphasized that the distribution of the lengths of the molecular chains in the product is not affected by the *rate* of reaction, but only by the *extent* to which it has proceeded as expressed by a value of p. But there are many step-growth polymerizations which would be expected to reach an equilibrium state (i.e. a balance between polymerization and depolymerization) when p is significantly less than 1.00 if the process were to be conducted in a closed system. But if these systems are conducted in a system which is open to the extent of allowing efficient removal of the coproduct, often water from the typical condensation reaction acts, high degrees of conversion can be achieved, i.e. $p \rightarrow 1.000$.

On other occasions it may be the case that polymerization needs to be stopped before the polymeric chains become too long for the application concerned (Section 3.2). One way is to quench the reaction mixture at the appropriate stage by rapid cooling: this relies upon the relatively high values of Arrhenius activation energies associated typically with the effective rate constant. A second procedure is to add a small amount of another reactant which has just one functional group per molecule. Once this has joined the chain with elimination of its single functionality, that end is unreactive thereafter. For example, acetic acid may be used in this way, terminating the growth of a chain through the reaction represented as

$$----\text{OH} + \text{HO}(\text{C}{=}\text{O})\text{CH}_3 \rightarrow ----\text{O}(\text{C}{=}\text{O})\text{CH}_3 \text{ (dead)}$$
$$+ \text{H}_2\text{O}.$$

5.2 Linear step-growth systems with two monomers

The two most important polymers produced industrially via step-growth mechanisms are poly(ethylene terephthalate) and nylon 6,6 (Table 1.1, Fig. 1.6), both of which are synthesized using two monomers. These monomers are bifunctional via two identical groups per molecule (Fig. 1.8) and may thus be represented as AA and BB following the notation used in the preceding section. The first elementary reaction which occurs is then represented as

$$\text{AA} + \text{BB} \rightarrow \text{AABB}$$

ignoring any small molecule eliminated in the action, which is condensation in these specific cases. Monomer molecules must join in strict alternation, so that the next two steps can be represented as

$$AABB + AA \rightarrow AABBAA$$

$$AABBAA + BB \rightarrow AABBAABB.$$

As far as general analysis with application of corresponding equations (5.1)–(5.8) is concerned, these systems with two monomers can be treated similarly to the system with the single AB monomer when the few following rules are applied

(i) the constitutional repeat units (CRUs), written as AABB (two monomer system) and AB (single monomer system), are regarded as equivalent;

(ii) only one of the functional groups (say A) is considered in assessing p, the fractional extent of reaction;

(iii) the AA/BB systems of interest have 1:1 monomer stoichiometries and the initial compositions have equal numbers of AA and BB molecules.

It is important at the outset to appreciate the significance of (iii) above. Consider a system in which the initial numbers of molecules of AA and BB are N_{AA} and N_{BB} respectively with $N_{AA} < N_{BB}$. A parameter r is equal to N_{AA}/N_{BB}. The extent of reaction (p) is defined in terms of the fraction of functional groups A which have reacted. Let ΔN be the number of groups A which have reacted at some time, when the corresponding p is expressed by

$$p = (2N_{AA} - \Delta N)/2N_{AA}. \qquad (5.9)$$

Substitution of $N_{AA} = rN_{BB}$ in the denominator leads to the equation

$$rp = (2N_{AA} - \Delta N)/2N_{BB}. \qquad (5.10)$$

The product rp measures the extent of reaction of the functional group B: stoichiometric considerations mean that ΔN is the number of B groups which have reacted and that $2N_{AA}$ is the limiting number of B groups which can react ultimately. The total number of unreacted functional groups at the ends of chains is expressed by

$$2N_{AA}(1 - p) + 2N_{BB}(1 - rp) = 2N_{AA}\{1 - p + (1 - rp)/r\}$$
$$= (2N_{AA}/r)\{r - rp + 1 - rp\}$$
$$= (2N_{AA}/r)\{1 + r - 2rp\}. \qquad (5.11)$$

The average degree of polymerization ($\overline{DP}$) is defined analogously to eqn (5.2) by

$$\overline{DP} = (2N_{AA} + 2N_{BB})/\{(2N_{AA}(1 - p) + 2N_{BB}(1 - rp)\}$$
$$= 2N_{AA}\{1 - (1/r)\}/\{(2N_{AA}/r)(1 + r - 2rp)\}$$
$$= (1 + r)/(1 + r - 2rp) \qquad (5.12)$$

with incorporation of eqn (5.11) and rearrangement. Equation (5.12) shows that as $p \rightarrow 1.000$, a limit is approached of $\overline{DP} \rightarrow (1 + r)/(1 - r)$. Table 5.3 lists these limiting values of $\overline{DP}$ for large extents of reaction when various excess fractional amounts of the monomer corresponding to BB are present initially. This makes it clear that even a 1 per cent excess of one of the monomers will shorten the average length of chains quite considerably. Thus to obtain relatively large polymer molecules, not only do almost exact stoichiometric proportions have to be achieved overall in the reaction vessel, but also there must be virtual elimination of poor mixing of the reactants on a local basis, the occurrence of side reactions and the presence of contaminants with some reactivity towards one of the functional groups.

In some step-growth polymerizations the average length of chains must be limited, as for example with nylon 6,6 for fibre production. This measure of control is commonly achieved by adding a suitable monofunctional agent, which could be represented as B in the scheme in which the bifunctional monomers are represented as AA and BB. An additional parameter is now defined as $r' = N_{AA}/(N_{BB} + N_B)$ when N_{AA}, N_{BB}, and N_B are the numbers of molecules of the subscript species. This equation recognizes that each B molecule is as effective in stopping a chain from growing as each additional molecule of BB which is present in excess of those required for $N_{AA} = N_{BB}$ the stoichiometric requirement. When a BB molecule joins onto a chain after all of the AA molecules have been consumed, the unreacted B functionality is unable to react further since there are no available A functionalities: this BB molecule terminates a chain

Table 5.3 Values of $\overline{DP}$ in a step-growth polymerization of two bifunctional monomers (AA and BB) for various fractional excesses of functionality B at various extents of polymerization (p)

p	Percentage excess of B	r	$\overline{DP}$
0.98	0	1.000	50
	0.1	0.999	47
	1	0.990	40
0.99	0	1.000	100
	0.1	0.999	95
	1	0.990	67
0.995	0	1.000	200
	0.1	0.999	182
	1	0.990	100
	2	0.980	68

just as effectively as would B. Thus the new parameter r' is simply substituted for r in eqn (5.12) in order to make quantitative predictions of $\overline{DP}$ in this type of system.

5.3 Some important linear step-growth polymers

This section outlines the syntheses of major polymers produced by step-growth processes operated on an industrial scale.

5.3.1 Nylon 6,6

The addition of exactly stoichiometric amounts (1:1) of the monomers initially is made easy in this case because these, adipic acid and hexamethylene diamine, are an acid and a base respectively. On mixing their solutions in methanol, a 1:1 salt is formed which precipitates out, being rather insoluble in this solvent. This salt could be termed hexamethylene diammonium adipate and is represented by the formula $H_3N^+(CH_2)_6N^+H_3$. $^-O(O=)C(CH_2)_4C(=O)O^-$. This is dissolved in water to give the concentrated solution added to the reaction vessel which is then sealed to provide an autoclave. Thus exactly equal numbers of molecules of the monomers in the form of salt ions are present as the autoclave is heated to approximately 550 K. As the temperature rises, a valve is opened periodically to allow the evolved steam to displace the air originally in the autoclave. Eventually the pressure is held at around 17 times atmospheric pressure before the steam is allowed to escape and finally drawn off by evacuation using a pumping system. The polymerization proceeds in the autoclave for typically 3–4 h. The addition of 0.5–1.0 per cent on a molar basis of monofunctional acetic acid (CH_3COOH) initially ensures that the chain lengths are limited. Thereafter nitrogen gas is injected under pressure to force the molten nylon 6,6 out through a valve or spinneret system at the bottom of the autoclave. The average degree of polymerization which is desirable when nylon 6,6 is intended for fibre production is in the range 40 to 70.

5.3.2 Poly(ethylene terephthalate) (PET)

The strategy used to synthesize polymeric chains of PET is very different to that used for nylon 6,6. The ethylene glycol monomer is not a strong enough base to give rise to a salt with the other, terephthalic acid, so that this route to exact stoichiometry is not available for PET synthesis. Rather PET synthesis starts with an excess of ethylene glycol and takes advantage of its relatively high volatility compared to other components of the reaction mixture. Ethylene glycol is drawn off above the mixture continuously as polymerization proceeds so that ultimately the monomer residues come towards the exact $1:1$ stoichiometric ratio.

In the most commonly used synthetic procedure, it is dimethyl terephthalate which is added to the reactor rather than terephthalic acid. The main reason is that the acid has only limited solubility in common solvents and also does not melt under practicable conditions, in fact subliming at around 574 K. The dimethyl ester is quite soluble and thus is reasonably easy to purify by recrystallization: moreover it is also volatile enough to be distilled. There are usually three stages in the industrial synthesis of PET, which have been termed melt transesterification, prepolymerization, and polycondensation.

The first stage takes place with excess ethylene glycol in the presence of a catalyst such as cobalt(II) or manganese(II) acetate. At the typical temperature of 470 K the overall result can be represented as

$$CH_3O(O=)C-C_6H_4-C(=O)OCH_3 + HOCH_2CH_2OH(xs)$$
$$\rightarrow HOCH_2CH_2O(O=)C-C_6H_4-C(=O)OCH_2CH_2OH$$
$$(+ \text{ some small oligomers}) + 2 \ CH_3OH\uparrow$$

where $-C_6H_4-$ represents the para- or 1,4-substituted aromatic ring. This process results typically in 80 to 90 per cent removal of the original methyl ester groups in the form of the drawn-off methanol.

The resultant mixture is injected into an autoclave for the subsequent 'prepolymerization' stage, conducted under reduced pressure (around 4 per cent of atmospheric pressure typically) at 530–560 K in the presence of a catalyst, commonly antimony oxide (Sb_2O_3). The main chemical process which is induced is a transesterification which can be represented as a condensation which eliminates ethylene glycol thus

$$2----C(=O)OCH_2CH_2OH \rightarrow$$
$$----C(=O)OCH_2CH_2OC(=O)---- + HOCH_2CH_2OH\uparrow.$$

Ethylene glycol is now essentially the only volatile substance involved and it can thus be drawn off to move the overall composition towards $1:1$ as chains join together. But even now the chain are relatively short and in bulk the mixture can only be termed a prepolymer.

The third stage takes place in a so-called finishing reactor which is operated under very reduced pressure (< 0.1 per cent of atmospheric pressure) to remove the most of the last remnants of the excess of ethylene glycol. It is common practice for this stage to be conducted with the PET as a melt, i.e. with the temperature above T_m (typically 538 K) and usually in the range 540–580 K. The reactor often has the basic design of a closed cylinder with its long axis horizontal. The average degree of polymerization which is achieved in the finished PET turns out to be governed mainly by the rate at which evolved ethylene glycol is extracted from the melt. To achieve this measure of control, another feature of the reactor system is an efficient agitator, often taking the form of a rotating screw, which not only maximizes the rate of release of ethylene glycol vapour from the rather viscous melt but also moves the polymerizing mass from the inlet at one end of the cylinder to the outlet at the other. This procedure is mainly used for producing PET of lower relative molecular mass, with $\bar{M}_n$ in the range 1.5–2.5×10^4 which is used to make fibres or films.

PET with higher $\bar{M}_n$ ($> 3.0 \times 10^4$) is required for injection or blow-moulding applications. The melt condensation route described above is less suitable for producing really high degrees of polymerization because thermal decomposition giving undesired side-products becomes significant at longer reaction times or higher temperatures. A solid-state polycondensation route has been developed to overcome these problems in the synthesis of PET with $\bar{M}_n > 2.0 \times 10^4$. This simply involves keeping the PET in the finishing reactor below T_m and accepting a longer timescale of reaction. To illustrate the beneficial effects, when the temperature was decreased from 558 to 503 K, the polycondensation rate decreased by a factor of six whereas thermal degradation slowed by around 40 times.

In melt-spinning of PET, the molten polymer is forced through the fine holes of a spinneret at the base of the finishing reactor, when it was cooled rapidly to temperatures below its T_g of approximately 340 K. As a result the filaments are almost totally amorphous: subsequent drawing with the temperature above 350 K induces the chain reorientation required to give the highly crystalline fibre (Section 3.2, Figs. 4.8, 4.9).

5.3.3 Nylon 6

Nylon 6 is a major commercial polymer, particularly in Europe. It is also referred to as polycaprolactam, on the basis of the common name of its single monomer more systematically termed 6-hexanolactam. In the presence of water (1–2 per cent) the overall polymerization process is described by the equation

$$n(CH_2)_5 \left\langle \begin{array}{c} C=O \\ | \\ N-H \end{array} \right. + H_2O \rightarrow$$

$$HO[-C(=O)(CH_2)_5N(H)]_{n-1}-C(=O)(CH_2)_5-NH_2.$$

Water has a catalytic role too, acting to open up the rings: in principle this action generates $HO-(CH_2)_5-NH_2$ units which can be considered as condensing together, regenerating water as the small molecule eliminated. Typically nylon 6 is synthesized in an autoclave at temperatures of around 540 K, water (steam) being drawn off as the process proceeds. In fact under the synthesis conditions the nylon 6 polymer is in equilibrium with a significant proportion of the monomer and this must be extracted by washing with water before the polymer can be spun. As is also the case with nylon 6,6 (Section 5.3.1) it is usually necessary to add small amounts of acetic acid, which adds onto the growing chains to give unreactive $CH_3C(=O)N(H)-$ or $CH_3C(=O)O-$ groups, thus limiting the lengths of the chains.

5.3.4 Silicones (polysiloxanes)

Figure 5.2 shows a flow diagram representing the synthetic route to poly(dimethylsiloxane) and ultimately a silicone rubber. The dichloro-silane $((CH_3)_2SiCl_2)$ is hydrolyzed to generate the unstable silanol $((CH_3)_2Si(OH)_2)$ which undergoes condensations to yield cyclic siloxanes, predominantly tri- and tetra-mers, under these conditions. The condensation to form a siloxane linkage can be represented as

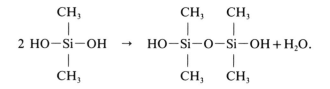

The principal cyclic siloxane used for preparation of polysiloxanes is the tetramer $((CH_3)_2SiO)_4)$ which is purified by distillation prior to ring-opening.

Silicone rubbers are produced typically via linear polymerization in the presence of an alkali metal hydroxide, KOH at ~410 K or NaOH some 30 K higher being the principal catalysts in this connection. The fillers used in producing the commercial forms of silicone rubber are usually forms of silica.

Silicone oils can be prepared by using sulphuric acid as the catalyst and adding a small amount of hexamethyldisiloxane $((CH_3)_3Si-O-Si(CH_3)_3)$. This results in the formation of polysiloxanes with inert, terminating trimethylsiloxyl $((CH_3)_3Si-O-)$ groups. The resultant silicone oils are thermally stable to quite high temperatures.

5.4 Network polymers

When at least one monomer with more than two reactive groups is a reactant in a step-growth polymerization, the polymeric products will have branched chains necessarily and often a continuous network structure is the result. An example of such a monomer is phenol which is trifunctional by way of being able to react at the 2-, 4-, and 6- positions on the aromatic

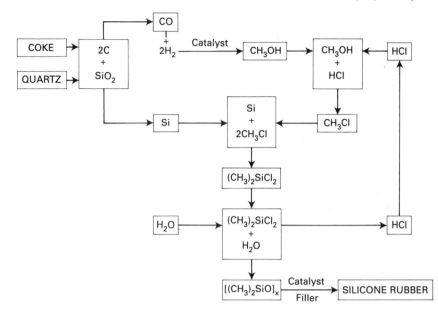

Fig. 5.2. Industrial synthetic route to polydimethylsiloxane and silicone rubber. Reproduced by permission of the Rubber Division, American Chemical Society, Inc., from Polmanteer, K. E. (1988). Silicone rubber, its development and technological progress. *Rubber Chemistry and Technology*, **61**, 470–502.

ring. Phenol–formaldehyde copolymers have network structures and are the major thermoset resins (Table 1.1 and Fig. 1.7).

The synthesis of phenol–formaldehyde polymers involves two stages usually. In the first, a *prepolymer* is produced, which might be described as a mixture of oligomeric or at most only partially polymerized molecules. The prepolymer has two important properties: it is effectively involatile and, even though viscous, it is a liquid which can be formed into the shape required. The subsequent stage completes the polymerization setting the hard rigid resin.

The commonest process uses a base as a catalyst in the first stage to yield a prepolymer known as a *resole*. Under these alkaline conditions, phenol is converted overwhelmingly to the resonance-stabilized anion, viewed conventionally as a mixture of the structures represented in Fig. 5.3(a). The carbon atoms with relatively high electronic density, indicated by two dots and a circled minus sign, are attacked by formaldehyde (HCHO) molecules to form methylolphenol species, effectively via addition and hydrogen atom transfer, as represented in Fig. 5.3(b) for the 2-position for example. These methylolphenol species remain susceptible to further attacks by formalde-

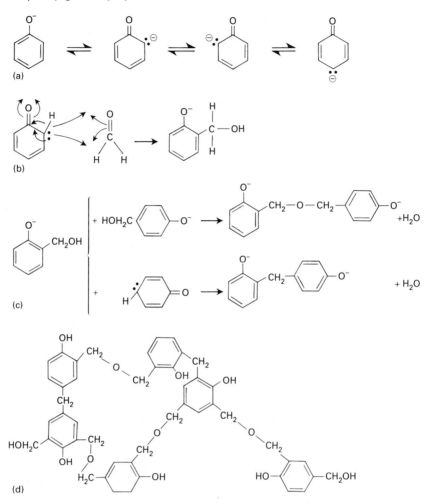

Fig. 5.3. Structures and mechanisms relevant to the synthesis of phenol–formaldehyde resole polymers.

hyde molecules at the other reactive positions. Also the methylol ($-CH_2OH$) groups can undergo mutual condensations, creating ether linkages ($-CH_2-O-CH_2-$) or they can condense at the functional positions of other rings, creating a methylene bridge ($-CH_2-$), as represented in Fig. 5.3(c). Further such actions build up prepolymer components of which a representative structure appears in Fig. 5.3(d): it may be noted that the linkages between aromatic rings are the result of the incorporation of either one ($-CH_2-$) or two ($-CH_2-O-CH_2-$) formaldehyde molecules.

In commercial practice resoles are usually produced by batch operations. In order, molten phenol is first into the reactor, followed by the catalyst, usually the hydroxide of sodium, calcium or barium. Formaldehyde in aqueous solution is then introduced at a rate which must be limited by the need to keep the temperature in the reactor close to 340 K when the exothermic reaction is taking place. Overall molar ratios of formaldehyde to phenol are usually in the range 1 to 3, reflecting the stoichiometry of the system. As the reaction proceeds water is taken out of the reactor in the form of liquid emerging from an attached condenser. Ultimately the resultant resoles will be set to the resin by subjecting the prepolymer material to temperatures around 425 K.

The main applications of resoles are in the manufacture of composite forms of wood, such as plywood or waferboard. For instance the prepolymer is mixed with wood flour and applied between layers of wood before the resultant sandwich is hot pressed, forming the resin bonding and steaming off evolved water. There is an optimal distribution of molecular weights which is governed by the diffusional rates of the prepolymer within the wood. If the molecules are too small, they will tend to diffuse too rapidly into the bulk of the wood, resulting in a 'starved' (i.e. poorly bonded) joint. Equally if the prepolymer molecules are too large, they will fail to penetrate into the wood sufficiently to produce strong bonding. Representative values for a satisfactory prepolymer in this connection are $\bar{M}_n \approx 700$ with a polydispersity $(\bar{M}_w/\bar{M}_n)$ of 4 to 4.5.

Another type of prepolymer can be obtained from the reaction of phenol and formaldehyde in the presence of an acidic catalyst, such as phosphoric or sulphuric acid. These *novolacs* (or *novolaks*) have higher values of $\bar{M}_n$ than resoles and usually come from a system containing an excess of phenol over formaldehyde with pH in the range 0.5 to 1.5. A major difference between novolac and resole prepolymers is that the former retain no methylol $(-CH_2OH)$ ring substituents as such (compare with Fig. 5.3(d)). As a consequence, a novolac cannot be polymerized on to a resin simply by raising the temperature, as is possible for resoles. Further addition of formaldehyde (or compounds such as paraformaldehyde which will generate formaldehyde by thermal decomposition) must be made to allow the resin to 'cure' (i.e. set rigid) at temperatures usually in the range 400–450 K. In the final resin from a novolac the linkages are predominantly methylene bridges $(-CH_2-)$ and there are few if any ether linkages $(-CH_2-O-CH_2-)$ between aromatic rings.

Urea–formaldehyde resins have already received attention through Fig. 4.5, Table 4.3, and the first review exercise of Chapter 4. As a monomer, urea $(H_2N-C(=O)-NH_2)$ is potentially tetrafunctional via the hydrogen atoms, which effectively add formaldehyde molecules to become methylol $(-CH_2OH)$ groups under mildly alkaline conditions. The evidence from ^{13}C NMR investigations (Table 4.3) suggests that the

resultant prepolymers have a relatively low degree of branching of the
chains which were predominantly terminated by $-CH_2OH$ groups. The
precise nature of the final resin is a function of both the ratio of the
feedrates of the monomers and the values of the pH at the various stages
of synthesis. No definitive structure can be specified however since there
is continuing debate on the microscopic nature of urea-formaldehyde
resins.

The main advantages of urea–formaldehyde resins over phenol–formaldehyde resins are that they are much lighter in colour and are generally
harder. But the darker phenolic resins offer better resistance to impact and
are also less susceptible to deterioration with exposure to moisture. Almost
three-quarters of the total production of urea–formaldehyde resins are in
fact used as bonding agents in the productions of composite woods and
furniture.

Melamine is a heterocyclic compound of aromatic nature which is derived
from cyanamide (Fig. 5.4(a)). It polymerizes with formaldehyde in much
the same way as does urea, first by creation of $-CH_2OH$ groups and then
by condensation of these with hydrogen atoms of $-NH_2$ groups to form
bridges (Fig. 5.4(b)). The resulting melamine–formaldehyde resins are
superior to the urea–formaldehyde resins, especially in terms of durability

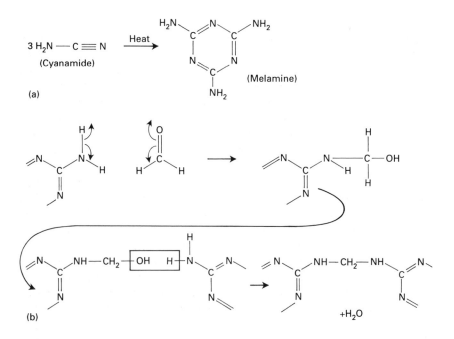

Fig. 5.4. Structures and mechanisms relevant to the synthesis of melamine–formaldehyde polymers (resins).

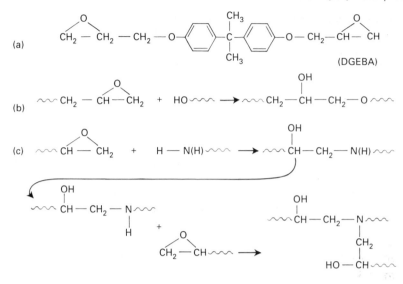

Fig. 5.5. Structures and mechanisms relevant to the formation of an epoxy resin from DGEBA and bisphenol A.

and the ability to retain a polished finish in normal usage. But the relatively high cost of production has restricted melamine-based resin to specialized usages, such as substituting for china in tableware and in laminating kitchen surfaces. The surfacing material known by the trade name of Formica is a several-layered sandwich of sheets of resin-impregnated 'paper'. The lowest layer, in which the resin is usually phenolic, contributes strength. The middle layer is a paper bearing the decoration which will be apparent through the pure cellulose layer above it in the finished product. Both of these layers are impregnated with melamine prepolymer initially. In the course of forming Formica the sheets of impregnated 'paper' are pressed together between heated plates of stainless steel, part of a press rated at five to eight million kilograms typically. The heat 'cures' the laminate, inducing the resin-forming polymerizations in the different layers.

Another type of network polymer is known generally as epoxy resin. The prepolymers are synthesized by step-growth mechanisms in systems where usually one monomer is a diepoxide and the other is a dihydroxy molecule. The commonest diepoxide is the diglyceridyl ether of bisphenol A, referred to by the acronym DGEBA, the molecular structure of which is shown in Fig. 5.5(a). The commonly used dihydroxylic species is bisphenol A. The chemical step represented in Fig. 5.5(b) occurs many times to create polyether linkages in a prepolymer of relatively low $\bar{M}_n$ (1000–3000 typically) with chains terminated by epoxide groups. In bulk the prepolymer is often a viscous liquid which is then mixed with the so-called 'curing agent' each molecule of which possesses several $-NH_2$ groups, for instance diethylene tetramine. Both hydrogen atoms of these $-NH_2$ groups have the ability to react with an epoxide ring, often at ambient temperature, to create β-hydroxyamino linkages between prepolymer molecules as represented in Fig. 5.5(c). This type of structure eventually extends to the gigantic network structure of the resin.

These types of polymeric materials are commonly available under various trade names, such as 'Araldite'. Epoxy resins are used widely as the bonding agents of fibreglass, as protective coatings for other materials, and as substances for sealing or encapsulating. In fact just over half of all the epoxy resin produced in Western Europe is used to produce coatings.

Exercises

5.1. Estimate the percentage error involved in assuming that the total mass of all monomer residues is $N_0 M$, where N_0 the initial number of molecules of the single monomer, $HO(CH_2)_{17}C(=O)OH$, and M is the relative molecular mass of the corresponding monomer residue in the polyester when the average degree of polymerization is 50.5.

5.2. Show that the chain length (x) for which w_x, the weight fraction as specified by eqn (5.5), has its maximum value is approximately equal to $1/p$ for $p = 0.90$, 0.95, and 0.98, where p is the extent of polymerization via a step-growth mechanism.

5.3. Poly(ethylene terephthalate) has been synthesized via melt polymerization to give a number average relative molecular mass $(\bar{M}_n)$ of 1.832×10^4. This was polymerized further in the solid state under flowing helium to achieve $\bar{M}_n = 9.032 \times 10^4$. Assuming that all terminal groups of chains are $-CH_2CH_2OH$, what mass of ethylene glycol would be expected to be evolved during the solid state polymerization per kg of the initial poly(ethylene terephthalate)?

Source: Ravindranath, K. and Mashelkar, R. A. (1990). *Journal of Applied Polymer Science*, **39**, 1325–45.

5.4. A synthesis of nylon 6,6 was conducted in an autoclave to which 250 moles of the corresponding nylon salt was added together with one mole of acetic acid. Predict the value of of $\bar{M}_n$ of the nylon 6,6 produced when the extent of polymerization is $p = 0.990$. What would the value of $\bar{M}_n$ have been without the added acetic acid?

5.5. Polyesters with the CRU structure represented as

$$-C(=O)-(CH_2)_{10}-C(=O)-CH_2CH((CH_2)_{17}\,CH_3)CH_2-O-$$

have been synthesized by polytransesterification of 2-octadecyl-1,3-propane-diol and the diphenyl ester of dodecanoic acid, in which phenol is volatilized and removed. Gel permeation chromatography (GPC) was used to investigate the mass distributions, when extents of conversion exceeded 0.97. The relative molecular masses (designated as M_p) corresponding to peaks of the GPC traces for a series of samples were found to cover the range 4.8×10^4 to 6.5×10^4.

(i) Write an equation to represent the act of transesterification concerned here.

(ii) On the basis of the established relationship $M_p = (\bar{M}_w/\bar{M}_n)^{1/2}$ and eqn (5.8) for $p \approx 1.00$, evaluate the average degrees of polymerization $(\overline{DP})$ of the samples.

(iii) Show that $\overline{DP} \geq 100$ is impossible for monomer ratios less than 0.98 and evaluate the p required to give $\overline{DP} = 100$ for $r = 0.995$, where r is the monomer molar concentration ratio initially (see eqn (5.12)).

Source: Andruzzi, F. and Hvilstead, S. (1991). *Polymer*, **32**, 2294-9.

5.6. Prepolymers to novolac (phenol–formaldehyde) resins have been synthesized. Nine oligomers containing 2–12 aromatic rings have been separated and quantitatively analysed by supercritical fluid chromatography. Relative numbers of molecules (N_i(rel)) having relative molecular masses M_i were measured as shown below

M_i	200	306	412	518	624	730	836	942	1048
N_i(rel)(%)	20.1	19.1	17.0	15.0	12.4	7.9	4.9	2.6	1.0

(a) Assign the likely structures to these species.

(b) The content of residual phenol in the novolac prepolymer was 5.00 per cent by mass. Evaluate $\bar{M}_n$ for the total product.

Data reproduced by permission of Elsevier from Mori, S., Saito, T. and Takeuchi, M. (1989). *Journal of Chromatography*, **478**, 181-90.

5.7. Figure 5.6 shows the solid-state ^{13}C NMR spectrum of a novolac (phenol–

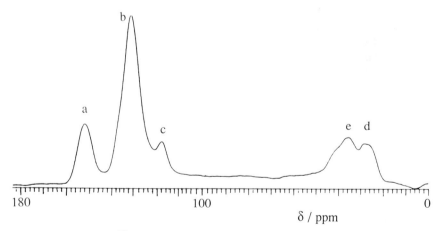

Fig. 5.6. Solid-state ^{13}C nuclear magnetic resonance spectrum obtained for a novolac (phenol–formaldehyde) resin prepared in the presence of sulphuric acid (with cross polarization, dipole decoupling, and magic angle spinning). Reprinted by permission of John Wiley & Sons, Inc., from Sinha, B. R., Blum, F. D., and O'Connor, D. (1989). Characterization of substituted phenol–formaldehyde resins using solid-state carbon-13 NMR. *Journal of Applied Polymer Science*, **38**, 163-171. Copyright (1989) John Wiley & Sons, Inc.

formaldehyde) resin. Assign the peaks labelled a,b,c,d, and e on the following bases:

(i) aromatic carbon atoms produce much larger chemical shifts than do aliphatic carbon atoms,

(ii) unsubstituted *ortho*-(2- or 6-) and unsubstituted *para*-(4-) carbon atoms give peaks that are coincident,

(iii) unsubstituted *meta*-(3- or 5-) and substituted *ortho* and *para*-carbons give peaks that are coincident.

Figure reproduced by permission of Wiley from Sinha, B. R., Blum, F. D. and O'Connor, D. (1989). Characterization of substituted phenol–formaldehyde resins using solid-state carbon-13 NMR. *Journal of Applied Polymer Science*, **38**, 163–71. Copyright (1989) Wiley.

5.8. Urea-formaldehyde resins in foam form are used to provide thermal insulation for buildings: over several years these materials deteriorate, becoming brittle and collapsing. Chemical procedures have been developed which result in the destruction of all $>N-CH_2-N<$ linkages in urea–formaldehyde and the generation of the same number of formaldehyde molecules. With a new dried foam, the yield of formaldehyde was 47 per cent on a weight for weight basis, but this fell to 33 per cent for an old foam. ^{13}C CP/MAS nuclear magnetic resonance spectra (Section 4.2) of these foams showed features assigned as follows with corresponding peak heights

Structure	Relative peak height	
	Fresh foam	Old foam
$-NH-{}^{*}CH_2-NH-$	59	70
$\begin{matrix} \vert \\ CH_2 \\ \vert \\ -N-{}^{*}CH_2-NH- \end{matrix}$	23	16
$\begin{matrix} \vert \quad\quad \vert \\ CH_2 \quad CH_2 \\ \vert \quad\quad \vert \\ -N-{}^{*}CH_2-N- \end{matrix}$	21	14
$>N-{}^{*}CH_2OH$	16	8

(*C is the carbon atom generating the signal.)

(i) What would be the expected yield by weight of formaldehyde released by the chemical procedure above if the polymer were composed entirely of linear chains represented as

$$H-[N(H)-C(=O)-N(H)-CH_2-]_n-N(H)-C(=O)-NH_2.$$

(ii) Deduce the general nature of the chemical processes which occur as urea–formaldehyde foams age.

6
Addition polymerization via free radicals

Vinyl polymers derived from monomers of the general formula $CH_2=CH(X)$ are produced in the highest volume of all types of polymer (Table 1.1, Fig. 1.4). Major thermoplastics, polyethylene (X=H), poly(vinyl chloride) (X=Cl), and polystyrene (X=C_6H_5), are synthesized predominantly by processes involving free radicals. Synthetic rubbers are mainly produced through radical addition polymerization of dienes of general formula $CH_2=C(X)-CH=CH_2$.

6.1 General features

Vinyl and diene monomers polymerize mainly via *straight chain* reactions based on chain-carrying species which are the growing polymeric chains. Introductory points have been made in Section 1.4.

In polymerization of a vinyl monomer, $CH_2=CH(X)$ head-to-tail linkage to yield the chain structure represented by repeating $-CH_2-CH(X)-CH_2-CH(X)-$ units may be presumed to predominate in general. A substantial rearrangement of electronic density must occur as a monomer molecule adds on to the growing chain to become a vinyl monomer residue. It is evident that π-electron density in the double bond of the monomer must be relocated in forming the new carbon–carbon σ-type of bond linking the resultant new residue to the chain. This rearrangement is unlikely to occur without the influence of some activating centre which has significant ability to pull electron density about. This centre may be an uncharged species with an orbital with a vacancy for one electron (i.e. a (free) radical centre) or for a pair of electrons (a coordinative vacancy) or it may be a species bearing a net electrical charge, either cationic (positive charge) or anionic (negative charge). Figure 6.1 represents a reaction act by which a chain grows by one monomer residue, i.e. a *propagation step* for each of these cases in the most general terms. In general, almost all vinyl and diene monomers can be polymerized by both ionic and radical routes, but usually one of these can be described as 'preferred', simply because the rate of polymerization in this case is much larger than the rates via the other mechanisms. The common exceptions are

Radical addition

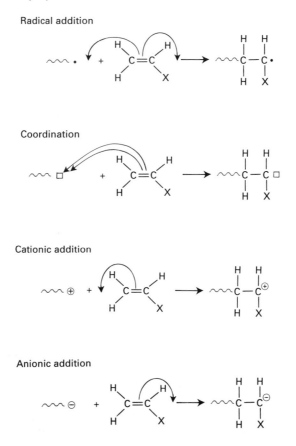

Coordination

Cationic addition

Anionic addition

Fig. 6.1. Representations of the propagation steps for the various types of addition polymerization mechanisms for the vinyl monomer $CH_2 = CH(X)$. Arrows show critical shifts of electron density, · represents an unpaired electron, and □ represents a vacancy for a pair of electrons.

polymerizations of monomers with conjugated systems of bonds, such as 1,3-butadiene and styrene, when polymerization occurs readily by radical and both ionic mechanisms. Otherwise the 'preferred' mechanism can normally be predicted by considering the effect of the group X on electron density. When X is a strongly electron-withdrawing group, a partial positive charge would be expected to be induced in the C=C bond: this would be expected to attract a growing chain bearing a negative charge, so it is perhaps no surprise that anionic polymerization tends to be favoured. Viewed in another way, $CH_2 = CH(X)$ with electron-attracting X is considered to be able to stabilize a carbanionic centre through induction

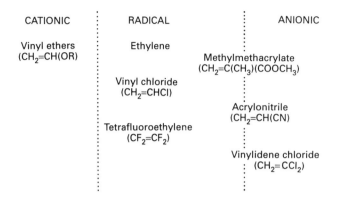

CATIONIC : RADICAL : ANIONIC

Vinyl ethers : Ethylene :
$(CH_2=CH(OR)$: : Methylmethacrylate
: : $(CH_2=C(CH_3)(COOCH_3))$
: Vinyl chloride :
: $(CH_2=CHCl)$:
: : Acrylonitrile
: : $(CH_2=CH(CN))$
: Tetrafluoroethylene :
: $(CF_2=CF_2)$:
: : Vinylidene chloride
: : $(CH_2=CCl_2)$

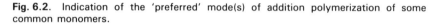

Fig. 6.2. Indication of the 'preferred' mode(s) of addition polymerization of some common monomers.

(or resonance) when this is formed by addition to an anionic chain. The reverse considerations apply for cationic polymerization to be favoured. As would be expected, when X is insufficiently strongly attracting or repelling for electron density, radical addition polymerization tends to be preferred. Figure 6.2 indicates the preferred mechanism(s) for some common monomers.

In fact, a much wider variety of monomers can be polymerized conveniently via free radical mechanisms than via ionic mechanisms. Furthermore radical centres can achieve polymerization in systems of very different physical natures, such as bulk liquids, homogeneous solutions, suspensions, and emulsions. On the other hand, ionic polymerizations are confined to homogeneous solutions: these and the even more restrictive coordination polymerization systems are the topics of the next chapter.

6.2 Physical forms of synthesis media

The interests in this section are the general features of the systems in common use for synthesizing polymers on large scales via radical addition mechanisms. These are considered in turn, giving some specific examples.

6.2.1 Bulk liquid monomers

Although the simplest system might seem to be that in which a pure liquid monomer polymerizes via the agency of free radicals generated within its bulk, this is not widely used because it is difficult to control. The main problem is the maintenance of a stable temperature throughout, when the polymerization reaction is strongly exothermic and the liquid becomes more viscous and thus less amenable to efficient heat extraction as polymerization proceeds. Furthermore the positive temperature coefficient of the rate constants (Table 6.1) means that localized rises of temperature in 'hot spots' will result in autoacceleration therein and thus significant inhomogeneity in terms of the conditions under which polymer chains were synthesized. Bulk liquid polymerizations are perhaps most important for laboratory studies, when suitable control can be imposed on the small volumes concerned.

6.2.2 Homogeneous solutions

In principle the dilution of the polymerizing system by dissolving it in an inert solvent would appear to solve the problems associated with polymerization of bulk liquid monomers. In particular the solution will never become too viscous for the efficient extraction of evolved heat. Later however there may be a difficulty in removing the solvent as fully as is required for many of the potential uses of the polymer. Hence the main commercial instances of solution polymerization via free radicals are when it is a solution of the resultant polymer which is the desired end product. This is the case in connection with the composition of many adhesives and paints. A specific case is poly(vinyl acetate), when this is to be hydrolyzed in solution to yield poly(vinyl alcohol). Moreover it is no surprise to find that solution polymerization is a commercial route to poly(tetrafluoroethylene) (PTFE); PTFE has no known solvent and thus precipitates out as fast as it is formed.

6.2.3 Suspensions

A suspension of a monomer is produced in the continuous phase of a liquid in which it is insoluble by steady mechanical agitation. Water is the usual liquid and a water-soluble stabilizer, such as poly(vinyl alcohol) or methyl cellulose, is usually required. The droplets of monomer which form the suspension are typically of diameters in the range 100–5000 μm. As the polymerization goes on, these droplets tend to become beads of solid polymer, which can be of a fairly uniform size if the suspension has been properly stabilized.

Suspension polymerization is induced to take place by the generation of free radicals *within* the monomer-droplets, i.e. in the discontinuous organic phase rather than in the surrounding aqueous phase. The usual means is through the thermal decomposition of a labile compound dissolved in the monomer. This localization of the source of free radicals within the organic phase is a main difference between suspension and emulsion polymerizations in common usage.

A major advantage of suspension polymerization is that temperature control is relatively easy. The large heat capacity of the water and the inability of small droplets to develop hot spots ensures efficient dispersal of evolved heat.

Most poly(vinyl chloride) is made in this type of polymerization system. Suspension polymerization is also used to produce hard glassy materials like polystyrene, polyacrylonitrile, and poly(methylmethacrylate) on industrial scales. Perhaps the main limitation is that the technique cannot be used to produce synthetic rubbers: the inherently sticky nature of these as synthesized would result in the coagulation of suspended droplets into large blobs which would then become stuck to stirrer blades, etc.

6.2.4 Emulsions

As with a suspension, the two phases of an emulsion polymerization system are immiscible liquids, but in the emulsion the (micro)droplet size is much smaller, typically of the order of $0.05-10\,\mu$m. The incorporation of a surfactant, such as a soap or detergent, is required to stabilize the emulsion. Usually the initiator system is located in the aqueous phase and the free radicals to which it gives rise cross into the organic droplets where they initiate polymerization. It is then apparent that emulsion polymerization is a heterogeneous process, with active species crossing phase boundaries.

A soap, such as sodium lauryl sulphate ($CH_3(CH_2)_9CH_2OSO_3^-\ Na^+$), or a detergent is an essential component of an emulsion polymerization system. At the outset it gives rise to *micelles*, each of which incorporates of the order of 100 soap molecules (Fig. 6.3(a)). The monomer resides mainly in other droplets of relatively large size ($\geqslant 1\,\mu$m diameter), which act as reservoirs of monomer rather than as reaction volumes: part of one is shown at the bottom right of Fig. 6.3(a).

A small fraction of the monomer molecules (M) enter the micelles, within which polymer (P) chains are synthesized. Free radicals (R·) coming in from the aqueous phase are able to penetrate the soap 'skin' of the micelles to encounter M molecules and start polymerization. The resulting growth of a chain is represented in Fig. 6.3(b) by the tangle of connected lines. As P forms, more M enters the micelle from the large droplet via diffusion of M molecules through the aqueous phase; these are represented by the isolated M symbols in parts (a), (b), and (c) of Fig. 6.3. Thus the large (reservoir) droplets shrink as time passes in a batch reaction system, whilst the micelles swell as they accumulate polymer chains, which is evident on comparing parts (a) and (c) of Fig. 6.3. A key point is that the number of micelles per unit volume of the aqueous phase far exceeds (typically by a

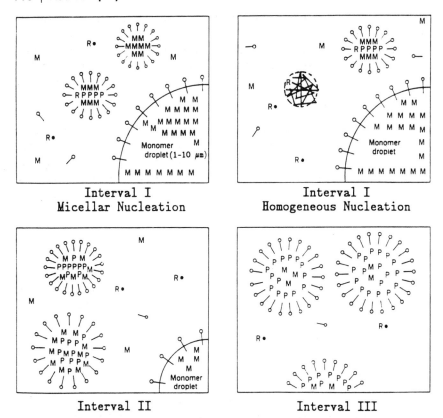

Fig. 6.3. A schematic representation of the microscopic course of events in a batch emulsion polymerization. R· represents an initiator radical, M a monomer molecule, P a polymer molecule, and O— a surfactant molecule. Interval I is the particle formation period, interval II is the constant growth period, and interval III is the period in which monomer concentration decreases sharply. Reproduced by permission of Hüthig & Wepf Verlag, Basel from Lee, D. I. (1990). The control of structure in emulsion polymerization. *Makromolekulare Chemie, Macromolecular Symposia*, **33**, 117–31.

factor of 10^7–10^8) the number of large droplets. As a result R· is far more likely to encounter a micelle than a large droplet as it diffuses through the water. In a typical emulsion polymerization system, the large droplets will have disappeared when 50–75 per cent of the initial M molecules have gone into P chains. Thereafter, at the stage represented in Fig. 6.3(d), the residual M are within polymer-swollen micelles and polymerization continues but much more slowly.

Emulsion polymerization is particularly suitable for the syntheses of rubbers which are the basis of elastomers. This is because the extremely small micelles in which the polymer is formed have soap skins and thus do not stick together as they would in the corresponding suspension system. Many of the most important synthetic rubbers originate in emulsion polymerizations: the list includes poly(chloroprene) (or Neoprene, the usual trade name) and also the copolymers known as SBR (styrene–butadiene rubber) and nitrile rubber (butadiene–acrylonitrile copolymer). Also the initial stage of producing ABS rubbers and plastics is the emulsion polymerization of butadiene.

It is of interest to find that Teflon or PTFE which is poly(tetrafluoroethylene) is produced in two very different forms via suspension and emulsion polymerizations. The former yields granular polymer whilst the latter produces a fine powder usually, as would be expected on the basis of the droplet sizes concerned. Commercially produced PTFE has very long chains, as indicated by typical values of $\bar{M}_w$ in the range 10^7 to 10^8, which are largely responsible for the intractable nature of this polymer in bulk.

6.3 Initiation processes

The act considered in this section is the first appearance of a reactive species with an unpaired electron but no electrical charge (i.e. a free radical) in addition polymerization systems. The different chemical means of effecting this are discussed in the following subsections.

6.3.1 Thermal decomposition of molecules

The commonest initiators under this heading have weak bonds which decompose at significant rates at moderate temperatures to yield free radicals. Quantitatively the usual kinetic requirement of such a thermal initiator is that it should decompose with a first-order rate constant of the order of 10^{-4}–10^{-6} s^{-1}, corresponding to half-lives of the order of 2–200 h. The usual weak bonds concerned are peroxide ($-O-O-$) or azo ($-N=N-$) ones.

α,α'-azobisisobutyronitrile (AIBN) is often used as a thermal initiator in radical addition polymerization systems. Its effective decomposition process is

$$(CH_3)_2C(CN)-N=N-C(CN)(CH_3)_2 \rightarrow 2\ (CH_3)_2C(CN)\cdot + N_2\uparrow$$

Nitrogen gas is evolved and cyanopropyl radicals are generated, which add on to a monomer molecule subsequently to commence the growth of the chain, i.e. propagation. AIBN is usually employed as thermal initiator at temperatures in the range 315–335 K, the first-order rate constant being around 1.3×10^{-5} s^{-1} at 335 K. Not every act of decomposition of an AIBN molecule leads to a pair of free radicals which are able to react with the monomer concerned. Some of the radicals fail to escape from the surrounding 'solvent cage' before they undergo other deactivating processes. Thus an *efficiency factor* (*f*) is defined as the apparent fraction of molecular decompositions which actually release free radicals so that they can react with monomer molecules in the solution. The actual initiation rate (R_i) of polymerization is the rate of decomposition of the initiator molecules modified by *f* and the number of radicals generated per molecule decomposed. The value of *f* for AIBN as a thermal initiator in solution is typically about 0.7.

Benzoyl peroxide is the peroxide most widely used as a thermal initiator. Its primary decomposition is represented by

$$C_6H_5-C(=O)-O-O-C(=O)-C_6H_5 \rightarrow 2\ C_6H_5-C(=O)-O\cdot$$

The typical temperature range in which it is suitable is 330 K ($\sim 9 \times 10^{-7} s^{-1}$) to 350K ($\sim 2 \times 10^{-5}\ s^{-1}$), the values of the rate contant at the limits being indicated in brackets. The primary (benzoyloxy) radical is subject to decomposition according to

$$C_6H_5-C(=O)-O\cdot \rightarrow C_6H_5\cdot + CO_2\uparrow$$

with a rate constant of $\sim 10^{-4}\ s^{-1}$ at 330 K. When they are first generated, the free radicals are in a solvent cage for some 10^{-9} s, a time short enough that decomposition to phenyl ($C_6H_5\cdot$) radicals within this cage is not significant. But beyond the confines of the cage, polymerization is initiated by both $C_6H_5COO\cdot$ and $C_6H_5\cdot$ radicals under normal conditions.

The organic initiators above are hardly soluble in water and are thus unsuitable for initiating emulsion polymerization. The useful species in this respect is the persulphate ion ($S_2O_8^{2-}$) which decomposes to yield two sulphate radical ions in a process represented as

$$S_2O_8^{2-}(aq) \rightarrow 2\ SO_4\cdot^-(aq)$$

The first order rate constant for this has a value of $\sim 2 \times 10^{-6}\ s^{-1}$ at 320 K and $\sim 2 \times 10^{-5}\ s^{-1}$ at 340 K, temperatures which limit the usual range of usage of potassium persulphate as an initiator in water.

6.3.2. Photodissociation of molecules

When light is absorbed by a molecule and causes it to break up (into radicals in the present context), this is termed *photodissociation*. The advantage of this mode of generating radicals for initiating polymerization is that the temperature of the system can be varied relatively widely, since the initiation rate is independent of it. Moreover the polymerization can be stopped quickly at any stage simply by switching off the light source. The rate at which radicals are generated and hence the concentration of growing chains depend on the intensity of the light. AIBN and benzoyl peroxide are common photoinitiators using near ultraviolet light, commonly the mercury line at a wavelength of 366 nm.

Only with emulsion polymerizations is photodissociative initiation of limited application, because most of the suitable molecules are virtually insoluble in water. In general, the method is only important on the laboratory rather than on the commercial scale.

6.3.3 Redox reactions

A *redox reaction* in which one (or more) electron(s) is transferred between reactant species per molecule reacted can be a particularly useful mode of initiation. There is also the advantage that most of the redox reactants in common usage are soluble in water. For example, one redox reaction which is used frequently to initiate emulsion polymerizations has as its reactants persulphate and thiosulphate ions and is represented as

$$S_2O_8^{2-} + S_2O_3^{2-} \rightarrow S_2O_3 \cdot^- + SO_4^{2-} + SO_4 \cdot^-$$

It is the sulphate radical ion ($SO_4 \cdot^-$) which is the active initiator. In a representative instance, the emulsion polymerization yielding a styrene-butadiene copolymer (for SBR elastomer production) was conducted in a batch system at 323 K which was operated for roughly 12 h to produce about two-thirds of the theoretical yield. Initially potassium persulphate composed 0.10 per cent of the emulsion, water 64 per cent, butadiene 24 per cent, styrene 10 per cent, and soap 1.7 per cent, all by weight.

Another redox initiation system involves the reaction of the ceric (Ce(IV)) ion, added in the form of ceric ammonium nitrate, with isopropanol ($i-C_3H_7OH$). The result of this reaction is summarized in general terms by stating that the reaction of Ce(IV) with the alcohol forms a complex, which subsequently decomposes to yield an organic radical, the active initiating species, together with Ce(III) and H^+ (aq). This system has, for example, been used to initiate the polymerization of methylmethacrylate in its aqueous solutions at temperatures around 310 K.

6.4 The growth and termination of chains

Once initiated, a chain will grow by repeated additions of monomer molecules with simultaneous recreation of the radical site at the new end. Propagating chains are highly reactive species in most radical addition polymerization systems. The act of propagation can be represented as

$$----\cdot + M \rightarrow ----M\cdot$$

where M represents the monomer; a typical order of magnitude of the rate constant, denoted by k_p, for this step is $10^3 \, dm^3 \, mol^{-1} \, s^{-1}$ (see Table 6.1). When the monomer concentration is representatively $[M] = 1 \, mol \, dm^{-3}$, the pseudo-first-order rate constant for propagation is $k_p[M] \sim 10^3 \, s^{-1}$, indicating that the average chain adds on a monomer residue on a time scale of 10^{-3} s or so and hundreds or even thousands will be added every second. Thus in evident contrast to step-growth polymerizations (Section 5.1), very long polymer chains are present in the earliest stages of addition polymerization.

The propagating chain will continue to grow until some process robs it of its free radical site, an act of *termination*. A common terminating mode is when two growing chains meet, combining their unpaired electrons to give a new backbone bond in a 'dead' polymer molecule or dispropor-tionating (often via transfer of a hydrogen atom) to give two 'dead' polymer molecules. This is often referred to as *mutual* termination of chains. Imagine the view of the polymerization system seen by a representative propagating chain. Monomer molecules are highly abundant whereas other growing chains are rare, being present at a very low concentration (typically of the order of 10^{-8} mol dm^{-3}) because of their reactive nature. It is therefore much more probable that this propagating chain will encounter and add on a monomer molecule (i.e. propagate) than meet another growing chain and undergo an act of mutual termination. This accounts in general for the relatively large resultant values of $\bar{M}_n$, ranging up to 10^7 or so, achieved in the systems unless the growing chains are intercepted in some other way.

It is clear that a reaction kinetic analysis must be made in order to understand how the average degree of polymerization $\overline{(DP)}$ changes as the conditions of the system are varied. Consider the simple mechanism, in which I is the initiating molecule, R· is the actual initiating radical, f is a factor corresponding to the efficiency with which R· radicals escape from the 'solvent cages' in which they are created, M is the monomer, and $----\cdot$ represents any growing chain.

$$
\begin{array}{lllll}
I & & \rightarrow & 2f\,\text{R}\cdot & \text{(initiation)} \\
\text{R}\cdot & +\ \text{M} & \rightarrow & \text{RM}\cdot & \\
\text{RM}\cdot & +\ \text{M} & \rightarrow & \text{RMM}\cdot & \text{(propagation)} \\
\end{array}
$$

$$
\begin{array}{lllll}
& & \overset{k_p}{} & & \\
----\cdot & +\ \text{M} & \overset{k_p}{\rightarrow} & ----\text{M}\cdot & \\
& & \overset{k_p}{} & \text{dead} & \\
----\cdot & +\ ----\cdot & \rightarrow & \text{polymer} & \text{((mutual) termination).} \\
\end{array}
$$

The rate constant, k_p, is independent of the length of the growing chain ($----\cdot$) other than for species composed of just a few monomer residues. When polymerization is proceeding at a steady rate, the rate of initiation (r_i) must be equal to the rate of termination (r_t) in order to keep the concentration of all growing chains, denoted by $\Sigma[----\cdot]$, effectively constant. It is conventional to regard the rate constant associated with the initiation step as that (k_d) for the decomposition of the initiator

I, so that the rate of initiation is defined as $r_i = fk_d[I]$. Hence the equations are obtained for steady polymerization

$$r_i = r_t = k_t(\Sigma[----\cdot])^2$$

which rearranges to give

$$\Sigma[----\cdot] = (r_i/k_t)^{1/2}. \tag{6.1}$$

The rate of propagation (r_p) is identical with the rate of polymerization of M when the chains are long and is given by the equation, with substitution of eqn (6.1),

$$r_p = k_p(\Sigma[----\cdot])[M] = k_p(r_i/k_t)^{1/2}[M]. \tag{6.2}$$

The average degree of polymerization ($\overline{DP}$) is simply the ratio r_p/r_i and from eqn (6.2) this is derived as

$$\overline{DP} = k_p[M] \ (r_i/k_t)^{1/2} / r_i = k_p[M]/(r_ik_t)^{1/2}. \tag{6.3}$$

This equation indicates that the average lengths of polymer chains (and hence values of $\overline{M}_n$) decrease as the rate of initiation (r_i) increases. Faster initiation increases the concentration of growing chains, $\Sigma[----\cdot]$, which enhances the termination rate ($\propto(\Sigma[----\cdot])^2$) more than the propagation rate ($\propto\Sigma[----\cdot]$), leading to lower $\overline{DP} = r_p/r_t$.

The overall rate of polymerization (R) is expressed by eqn (6.4).

$$R = -d[M]/dt = r_p = k_p(\Sigma[----\cdot])[M]$$
$$= k_p(k_d/k_t)^{1/2} \cdot f^{1/2} \cdot [I]^{1/2} \cdot [M] \tag{6.4}$$

when eqn (6.1) and $r_i = fk_d[I]$ are substituted. For a large number of radical addition polymerizations conducted in the bulk liquid monomer, solution or suspension modes, the experimental rate law has the form

$$R = k_{obs}[I]^{1/2}[M]$$

which, in comparison with eqn (6.4), indicates the identification of the observed rate constant as

$$k_{obs} = f^{1/2} \cdot k_p(k_d/k_t)^{1/2}. \tag{6.5}$$

The variation with temperature of each of the rate constants is expected to be expressed by an Arrhenius form, for example $k_{obs} = A_{obs} \cdot \exp(-E_{obs}/RT)$, with apparent activation energies E_d, E_p, and E_t appearing in the corresponding equations for the rate constants k_d, k_p, and k_t for elementary reactions. The parameter f is almost insensitive to temperature in comparison. Converting eqn (6.5) into its logarithmic form and then differentiating with respect to reciprocal temperature yields the equalities

$$d(\ln k_{obs})/d(1/T) = -E_{obs}/RT$$
$$= d(\ln k_p)/d(1/T) + \tfrac{1}{2}d(\ln k_d)/d(1/T) - \tfrac{1}{2}d(\ln k_t)/d(1/T)$$
$$= -E_p/RT - \tfrac{1}{2}E_d/RT + \tfrac{1}{2}E_t/RT.$$

These lead to the equation

$$E_{obs} = E_p + \tfrac{1}{2}E_d - \tfrac{1}{2}E_t. \qquad (6.6)$$

When either of the common thermal initiators, AIBN or benzoyl peroxide, is used, $E_d \approx 126\,\text{kJ mol}^{-1}$, a value which corresponds closely to the strength of the bond which breaks. Table 6.1 lists some representative values of rate constants and activation energies for commercially important polymers, with E_{obs} deduced on the basis that $E_d = 126\,\text{kJ mol}^{-1}$ exactly.

Emulsion polymerization has kinetic features which are different from those of the other modes. Usually the initiating radical is generated in the continuous aqueous phase at rates of the order of 10^{16} radicals $\text{dm}^{-3}\text{s}^{-1}$. Typically there would be of the order of 10^{17} micelles dm^{-3} of emulsified organic phase (Section 6.2.4). On average a radical enters a micelle at a representative rate of one every 10 s. When a radical enters a micelle already containing a growing chain, these are expected to come together and mutually terminate on a much shorter timescale, say in milliseconds. Termination in the aqueous phase is not significant. This suggests that over a relatively long period of time it may be assumed that each micelle in the system contains zero or one growing chain with equal probability. Figure 6.4 can be viewed as a series of snapshots of an average micelle which is entered by a free radical from the surrounding aqueous phase at regular time intervals.

In the time intervals $0 \rightarrow \Delta t$, $2\Delta t \rightarrow 3\Delta t$, and $4\Delta t \rightarrow 5\Delta t$ the micelle contains no growing chain ($- - - - \cdot$), only dead polymer molecules (P). In the intervening time intervals ($\Delta t \rightarrow 2\Delta t$, $3\Delta t \rightarrow 4\Delta t$, $5\Delta t \rightarrow 6\Delta t$) the micelle contains one growing chain. This indicates the *ideal case assumption* that on average each micelle in the emulsion polymerization system contains

Table 6.1 Representative values of rate parameters for radical addition polymerizations with thermal initiation

Monomer	T (K)	k_p^*	$10^{-7}k_t^*$	$E_p^\dagger$	$E_t^\dagger$	$E_{obs}{}^\dagger$
Ethylene	355	240	54	18	1	80
Vinyl chloride	323	1100	21	16	18	70
Styrene	333	180	5.5	29	10	87
Tetrafluoroethylene	310	7200	7.3	17	14	73
Acrylonitrile	333	2000	78	16	16	71
Vinyl acetate	323	2700	8.0	35	14	91
Methylmethacrylate	333	520	2.6	26	12	83

$^*\text{dm}^3\,\text{mol}^{-1}\,\text{s}^{-1}$.
$\dagger\,\text{kJ mol}^{-1}$.

a growing chain for half of the time. The alternative view is that at any instant of time half of the micelles contain a single growing chain and the other half contain none.

Let N be the number of micelles in total emulsion volume of V, corresponding to an effective micelle concentration of N/V, and [M] is the monomer concentration within the micelles which can be expected to be the same as that in the pure liquid monomer. In the ideal system, the rate of propagation (r_p) is expected to be expressed as

$$r_p = \tfrac{1}{2} k_p [M] (N/V)/N_A$$

where k_p is the rate constant of the propagation step and N_A is the Avogadro number (6.02×10^{23}) and is required to convert N/V to the equivalent molar concentration. This system behaves kinetically as would another system composed only of the bulk liquid monomer containing a concentration of growing chains $\Sigma[-----\cdot] = \tfrac{1}{2} N/(N_A V)$. In the emulsion polymerization system, [M] is essentially constant for all conditions so that r_p and hence the overall polymerization rate are governed by the number of micelles present but *not* by the rate of initiation.

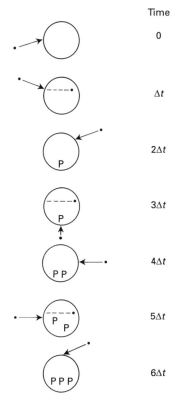

Time

0

Δt

$2\Delta t$

$3\Delta t$

$4\Delta t$

$5\Delta t$

$6\Delta t$

Fig. 6.4. Diagram representing a series of 'snapshots' of a micelle during emulsion polymerization at critical time intervals (Δt) between the entry of radicals $(\cdot)$ from the surrounding aqueous phase.

The rate at which the radicals are transferred from the aqueous phase into the micelles is assumed to be equal to the rate (ρ_i) of generation of these radicals throughout the aqueous phase. The rate at which chains are initiated in the micelles is then $\frac{1}{2}\rho_i$, since half of the radicals entering micelles terminate growing chains already in them. Hence, the average degree of polymerization in the ideal case of emulsion polymerization is expressed by the equation

$$\overline{DP} = r_p/\tfrac{1}{2}\rho_i = \tfrac{1}{2} \, k_p[M] \cdot \{N/(N_A V)\}/\tfrac{1}{2}\rho_i$$
$$= (k_p/\rho_i)[M] \cdot [\text{micelles}] \qquad\qquad (6.7)$$

where $[\text{micelles}] = N/(N_A V)$. Hence $\overline{DP}$ (and hence $\bar{M}_n$) of polymer synthesized by a radical addition polymerization mechanism in an emulsion together with the rate of polymerization will both be expected to increase with greater addition of emulsifying agent since this increases the number and hence effective concentration of micelles.

Several emulsion polymerization systems are known which show kinetic behaviour corresponding to the ideal case. But there are other emulsion polymerization systems which show marked deviations, particularly in showing kinetic behaviour which suggests that the average number of growing chains per micelle is less than 0.5.

6.5   The control of chain growth

Mutual termination steps involving two chains annihilating their active sites for growth may be considered as the 'normal' means of creating a dead polymer molecule. This may involve *combination* represented as

$$----CH_2CH(X)\cdot + \cdot CH(X)CH_2----$$
$$\rightarrow ----CH_2CH(X)-CH(X)CH_2----.$$

In this event the final polymer chains are on average twice the length of the growing chains. A particular example is provided by polystyryl radicals ($X = C_6H_5$) which are believed to terminate only by combination. Alternatively mutual termination may be via *disproportionation*, which forms two dead polymer molecules which retain the same number of monomer residues as were in the growing chains when they reacted together. This process involves the transfer of a hydrogen atom from one of the radicals to the other, as is represented below.

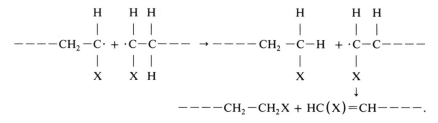

Poly(methylmethacrylate) synthesis provides a substantial instance, about two-thirds of termination being via disproportionation with the remainder being via combination of $----CH_2-C(CH_3)(C(=O)OCH_3)\cdot$.

An *inhibitor* is often defined in the present context as a substance which reacts so rapidly to eliminate the initiating radicals ($R\cdot$ in the mechanism above) and/or radical sites on very short chains that polymerization cannot start until all of the molecules of the inhibitor have been removed from the system. Inhibitors are often added to monomers in bulk to prevent unwanted polymerization, say initiated by cosmic rays, occurring during storage and transport. Then there is an essential separation of monomer from the inhibitor, say via distillation, prior to its introduction to the polymerization reaction system. Typical inhibitors are quinones, such as 1,4-benzoquinone.

In laboratory studies of the inhibition of radical addition polymerization it is often advantageous to use a stable free radical as inhibitor, usually diphenylpicrylhydrazyl (DPPH) with the structure $(C_6H_5)_2-N-\dot{N}-C_6H_2(NO_2)_3$, the aromatic ring on the right being substituted with nitro groups at the 2-, 4-, and 6-positions. The actual rate of initiation of polymerization can be measured using an inhibitor like DPPH. Figure 6.5 represents the effect of various additions of DPPH to a system in which methylmeth-

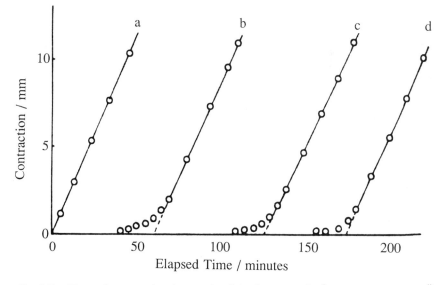

Fig. 6.5. Plots of contraction (proportional to the amount of monomer consumed) versus time for the polymerization of methylmethacrylate (MMA) in benzene solution at 303 K, when the added concentrations are 2.00 mol dm^{-3} of monomer (MMA) and 0.1 mol dm^{-3} of initiator (AIBN) with added concentrations of inhibitor (DPPH) of 0 (a), 0.456 (b), 0.798 (c), and 1.14 (d) $\times 10^{-4}$ mol dm^{-3}. Reproduced by permission of the Society of Polymer Science, Japan, from Kamachi, M., Liaw, D. J., and Nozakura, S-i. (1981). Solvent effect on radical polymerization of methylmethacrylate. *Polymer Journal*, **13**, 41–50.

acrylate (MMA) was being polymerized in solution in benzene using AIBN as the thermal initiator. The presence of the inhibitor leads to an *induction period*, a time lapse at the start before any polymerization occurs. Evidently during the induction period, DPPH is reacting with the initiating radicals much more rapidly than these radicals can react with monomer molecules, the effective action being the two steps

$$\text{AIBN} \quad \rightarrow \quad 2f\,\text{R}\cdot$$
$$\text{R}\cdot + \cdot\text{DPPH} \quad \rightarrow \quad \text{R}-\text{DPPH}.$$

There is a linear relationship between the added concentration of DPPH and the length of the induction period: the corresponding gradient is equal to the rate of generation of initiating radicals. For example, when $[\text{AIBN}] = 0.10\,\text{mol dm}^{-3}$ and $[\text{MMA}] = 2.00\,\text{mol dm}^{-3}$, the rate of generation of initiating radicals was $(1.09 \pm 0.04) \times 10^{-8}\,\text{mol dm}^{-3}\text{s}^{-1}$ at 303 K.

Chain transfer is an important phenomenon in many radical addition polymerization systems: it results in the termination of the growth of one polymer chain through the transfer of the free radical centre to another species. Thus this reduces the probability of longer chains and accordingly restricts the size of $\overline{\text{DP}}$ and $\bar{M}_n$ of the polymer consequently. A general representation of chain transfer can be given in terms of growing chains $(----\cdot)$ and molecules $(\text{X}-\text{Y})$

$$----\cdot + \text{X}-\text{Y} \rightarrow ----\text{X} + \text{Y}\cdot$$

Commonly the identity of X is a hydrogen atom (H). One point of interest is the role of chain transfer in preventing polypropylene being produced by radical addition polymerization. Chain transfer here is represented as

$$----\cdot + \text{H}-\text{CH}_2-\text{CH}=\text{CH}_2 \rightarrow ----\text{H} + \overbrace{\text{CH}_2=\text{CH}=\text{CH}_2}^{\cdot}$$

The product radical is allylic, which is resonance stabilized and therefore rather unreactive. This process is termed *degradative transfer*, when the radical resulting from transfer is unable to propagate. This type of transfer is a form of termination and is the reason that polypropylene with relatively high values of $\bar{M}_n$ cannot be synthesized in systems with free radicals.

The relative sizes of the rate constant for the transfer step (k_{tr}) and the rate constant for a propagation step (k_p) determine different situations. Inhibition, or more specifically *telomerization*, is said to occur when the ratio k_{tr}/k_p exceeds 10. The general mechanism is written

		I	$\rightarrow$	$2f\,\text{R}\cdot$	(1)
R·	+	M	$\rightarrow$	RM·	(2)
RM·	+	M	$\rightarrow$	RM$_2$·	(3)

....

$----\cdot$	+	M	$\xrightarrow{k_p}$	$----\text{M}\cdot$	
$----\cdot$	+	XY	$\xrightarrow{k_{tr}}$	$----\text{X} + \text{Y}\cdot.$	

If it is assumed that the rate constants for early steps such as (2) and (3) are not very different in size from k_p, it can be seen how telomerization stops any significant appearance of polymer molecules. Only a few per cent at most of $R\cdot$ becomes $RM\cdot$ and in turn a similar small proportion of $RM\cdot$ becomes $RM_2\cdot$. Even if the radical $Y\cdot$ resulting from chain transfer can propagate, chains are still short when they terminate. An example is 1,4-benzoquinone acting like XY in the radical polymerization of styrene initiated thermally by AIBN at 323 K. Here the value of k_{tr}/k_p is 520 so that even minor concentrations of this inhibitor will be effective in preventing propagation from occurring until all of it has been destroyed. On the other hand, in the same system with tetramethylbenzoquinone acting as XY, k_{tr}/k_p is only 4.4 and polymer formation is not stopped but only much slowed down; this substance is termed a retarder hence.

Perhaps the most important aspect of chain transfer from the point of view of the commercial syntheses of polymers via radical addition concerns *modifiers* which are chain transfer agents which allow the average degree of polymerization to be controlled or varied as required. Mercaptans (general formula RSH with R an alkyl group) are common modifiers, their chain transfer action being represented as

$$----\cdot + RSH \rightarrow ----H + RS\cdot.$$

Generally $RS\cdot$ is a radical which can initiate growth of a new chain. Modifiers usually transfer hydrogen atoms and so are conventionally denoted by TH in general terms.

Consider the steps which propagate, transfer, and terminate chains which are usually the set

$$----\cdot \ + \quad M \quad \xrightarrow{k_p} \ ----M\cdot \qquad\qquad \text{(propagation)}$$
$$----\cdot \ + \quad TH \quad \xrightarrow{k_{tr}} \ ----H \ + \ T\cdot \quad \text{(transfer)}$$
$$----\cdot \ + \ ----\cdot \ \xrightarrow{k_t} \ \text{dead polymer} \qquad \text{(termination).}$$

The average degree of polymerization $(\overline{DP})$ must now be redefined as the ratio of the propagation rate to the sum of the rates of termination and transfer, expressed as

$$\overline{DP} = \frac{k_p[M]\,\Sigma[----\cdot]}{k_t\,(\Sigma[----\cdot])^2 + k_{tr}\,[TH]\,\Sigma[----\cdot]}$$

$$= \frac{k_p[M]}{k_t\,\Sigma[----\cdot] + k_{tr}[TH]}.$$

It is assumed that k_{tr}, like k_p and k_t, is independent of the length of the chain concerned. Both sides of the equation above are inverted to give the equation

$$1/\overline{DP} = (k_t\,\Sigma[----\cdot]\,/\,k_p[M]) + (k_{tr}\,/\,k_p)\cdot[TH]/[M]$$

$$= 1/\overline{DP}_0 + (k_{tr}\,/\,k_p)\cdot[TH]/[M]$$

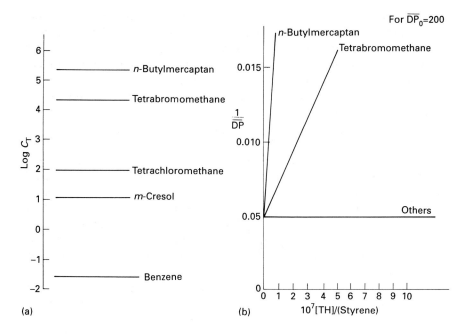

Fig. 6.6. (a) Values of chain transfer constants (C_T) of various substances in the polymerization of styrene at 333 K. (b) Corresponding plots of the reciprocal of the average degree of polymerization ($\overline{DP}$) versus the ratio of concentrations of the chain transfer agent (TH) and styrene for three different TH species (Mayo plots).

in which $\overline{DP}_0$ is the average degree of polymerization achieved when the modifier TH is not present. The rate constant ratio k_{tr}/k_p is often termed the *chain transfer constant* (C_T) for a specified system, so that the equation becomes

$$1/\overline{DP} = 1/\overline{DP}_0 + C_T \cdot [TH]/[M]. \qquad (6.8)$$

This is a form of what is sometimes known as the Mayo equation.

Figure 6.6(a) indicates the values of C_T for various identities of TH in the radical addition of styrene at 335 K and Fig. 6.6(b) shows corresponding linear plots predicted by eqn (6.8).

The dispersity of the resultant polymer can be specified in terms of a parameter S defined in terms of the rates (r) of the steps in the mechanism given above as $S = r_p/(r_p + r_{tr} + r_t)$. For simplicity consider only the case in which each growing chain yields one polymer molecule, i.e. when termination is via disproportionation rather than by combination. Imagine that there was a miniature device which could enter the polymerization system and attach itself at random to the residue (R) of an initiating radical at one end of a dead polymer chain. The probability that the monomer residue (M) directly attached to R is attached to another M (i.e. the probability that the chain end structure is RMM) is equal to S: during polymerization, S is the fraction of radical centres at any instant which propagate rather than become dead through termination or transfer. In fact S here is directly analogous to the parameter p of Section 5.1, so that, following the form of eqns (5.6) and (5.8), the relevant equations in the present connection are

$$\overline{DP} = 1/(1 - S) \ (a), \ \bar{M}_w/\bar{M}_n = 1 + S \ (b). \tag{6.9}$$

In a typical addition polymerization r_p must be considerably greater than $r_{tr} + r_t$, otherwise chains long enough to be described as polymers would not be formed. Hence $S \rightarrow 1.00$ and the polydispersity $\bar{M}_w/\bar{M}_n \rightarrow 2.00$.

Finally in this section it is essential to focus on the importance of chain transfer in many emulsion polymerization systems. In the ideal case only one growing chain can be present in a micelle of monomer and if k_p is relatively large, extremely long chains could result which would be unlikely to make the bulk polymer produced have the desired properties for general usage. A chain transfer agent (or modifier) is therefore usually added to limit the value of $\overline{DP}$ to what is needed. A typical formulation for the synthesis of styrene–butadiene copolymer (the basis of SBR elastomer) is shown in Table 6.2, where the roles of minor components are indicated. The overall kinetic behaviours of such emulsion polymerizations are sometimes difficult to model; one of the complications is often chain transfer to the emulsifying agent.

Table 6.2 Typical composition of polymerization medium (weight %) in the emulsion synthesis of styrene–butadiene copolymer

Water	64	Emulsifying agent (fatty acid soap) 1.7
Butadiene	24	Thermal initiator (potassium persulphate) 0.10
Styrene	10	Chain transfer agent (n-dodecyl mercaptan) 0.17

6.6 Generation of side chains

A polymer consisting of molecules which are all linear chains will usually have very different properties from the same basic polymer composed of molecules in which the backbone has pendant chains or is branched.

In some radical addition polymerization systems, the development of chain branching is an inherent feature and must be accepted. Branching can be the result of *internal chain transfer*, where the growing end of the chain has a tendency to abstract a hydrogen atom from a $-CH_2-$ group further back in the same chain, a process termed *backbiting*. This occurs in the high pressure polymerization of ethylene when the product is known

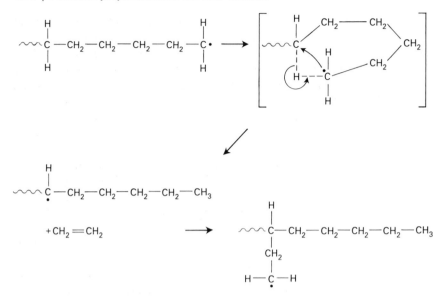

Fig. 6.7. Representation of a typical form of backbiting in the synthesis of polyethylene by radical addition polymerization.

as low density polyethylene (LDPE). The backbone of LDPE has many pendant chains composed of a small number of linked carbon atoms, typically five or six, and an illustration of the backbiting concerned is given in Fig. 6.7. This type of transfer is very significant for the properties of the final bulk polymer (see Section 8.1).

Side chains can be created on existing polymer molecules by *grafting*, in which new chemical bonds are created between atoms in the backbone and the precursors to the side chains. A *graft copolymer* is formed when chains of one polymer are attached along the length of another type of chain. Perhaps the pre-eminent graft copolymer from the commercial point of view is known commonly as ABS, the acronym of the three monomers, acrylonitrile, butadiene, and styrene. The first step in producing ABS copolymers is the polymerization of butadiene, often in an emulsion system. Free radical sites (i.e. unpaired electrons) are generated on the resultant polybutadiene chains; suitable chain transfer agents in conjunction with a persulphate initiator result in the formation of radicals which remove hydrogen atoms from carbon atoms in the backbone. When this is done in the presence of acrylonitrile and styrene, a random copolymer of these grows from the sites concerned to produce the grafted polymer. The content by weight of butadiene can be from 8 to 65 per cent, depending upon the properties of the ABS product required (see Section 9.1).

Another example of grafting by transfer is provided by the thermal decomposition of benzoyl peroxide in the presence of methylmethacrylate (MMA) and polystyrene. The radicals commence the polymerization of the MMA and some of the chains are terminated by transfer of hydrogen atoms from $-CH_2-$ components of the backbone of the polystyrene, leaving radical sites thereon. These also propagate, so that side chains of poly-(methylmethacrylate) grow out from the polystyrene chain. The grafted nature of the resultant copolymer is indicated by designating it as poly(styrene-g-methylmethacrylate).

Grafting is one of the best ways of achieving the 'mixing' of polymers which are effectively incompatible. Grafted copolymers have the properties of both separate polymers to some extent. For example poly(vinyl chloride) is not highly resistant to impact but it can be produced in a so-called 'high impact' form by grafting poly(vinyl chloride) chains onto polybutadiene chains, the latter having a rubbery nature. This copolymer would be more properly termed poly(butadiene-g-vinyl chloride). In one instance this grafting by transfer was accomplished in an reactor fed with polybutadiene ($\bar{M}_n \approx 4500$), vinyl chloride, and AIBN as initiator and operated at 333 K. The radicals from the decomposition of AIBN commenced polymerization of vinyl chloride and the resultant growing chains have a transfer constant (C_T) (eqn (6.8)) of 11 with respect to polybutadiene acting as a TH molecule. The basic chemical action in this transfer grafting may be represented simply as

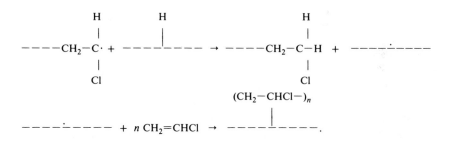

Grafting may also be accomplished using high-energy radiation, such as γ-rays or electron beams to damage polymer chains and thus create propagation sites for a monomer along their lengths. A large amount of research work has been performed in laboratories over the last few decades but as yet there does not seem to have been any large-scale commercial production of radiation-grafted polymers.

Discussion of the possible future use of macromonomers and iniferters in the synthesis of graft copolymers is reserved till Section 10.1.

Exercises

6.1. What is the 'cage effect' for reactions in liquid phases and what are 'geminate' reactions? (Look these up in an appropriate textbook on reaction kinetics.) Hence explain why you would expect the value of the efficiency factor (f) in thermal initiation (Section 6.3.1) to be less than 1.00 in general.

6.2. Estimate the value of f applicable to the system shown in Fig. 6.5 on the basis that the first-order rate constant for the decomposition of the initiator (AIBN) is $k_d = 1.52 \times 10^{-7} s^{-1}$ at 303 K.

6.3. In the initial stages of polymerization of pure liquid methylmethacrylate at 333 K using AIBN as thermal initiator, electron spin resonance (ESR) spectroscopy was used to measure a *total* radical concentration of $1.5 \times 10^{-7} mol\,dm^{-3}$. Use $k_d = 9.9 \times 10^{-6} s^{-1}$ for the thermal decomposition of AIBN and the value of k_t given in Table 6.1 to evaluate f when [AIBN] = 0.095 mol dm^{-3}.
Source: Garrett, R. W., Hill, D. J. T., O'Donnell, J. H., Pomery, P. J., and Winzor, C. L., (1989). *Polymer Bulletin*, **22** 611–16.

6.4. Potassium perphosphate ($K_4P_2O_8$) and mercaptosuccininic acid (RSH) provide a redox initiator system through the reaction

$$P_2O_8^{4-} + RSH \rightarrow RS\cdot + HPO_4^{2-} + PO_4\cdot^{2-}$$

This system has been used to initiate the polymerization of acrylamide (M) in aqueous solution. The experimental rate law was first order in M and half order in each of $K_4P_2O_8$ and RSH. How may this be interpreted?
Source: Behari, K., Raja, G. D., and Agarwal, A. (1989). *Polymer*, **30**, 726–31.

6.5. Styrene has been polymerized in an aqueous emulsion at 333 K. The density of particles (micelles) was $9.41 \times 10^{16} dm^{-3}$ when the rate of polymerization was $1.23 \times 10^{-4} mol\,dm^{-3}\,s^{-1}$. The density of pure liquid styrene is $0.906\,kg\,dm^{-3}$ and its relative molecular mass is 104. Using Table 6.1 as the source of kinetic data, show that this corresponds to an ideal emulsion polymerization. If $\bar{M}_n$ of the polystyrene is 9.76×10^4, what is the rate of transfer of free radicals from the aqueous phase into the micelles?

6.6. Methylmethacrylate (MMA) has been polymerized in bulk at 333 K using AIBN as thermal initiator. Ethyl(tertbutylthio)methylacrylate (EBTMA) ($CH_2{=}C(C({=}O)OCH_2CH_3)CH_2SC(CH_3)_3$) was used as the modifier (chain transfer agent). The following data were obtained for the average degree of polymerization ($\overline{DP}$) as a function of the concentration ratio of modifier to monomer

10^3 [EBTMA]/[MMA]	0	1.28	2.37	4.52
$\overline{DP}$	1367	618	408	245

What is the value of the chain transfer coefficient?
Source: Miejs, G. F., Rizzardo, E., and Thang, S. H. (1988). *Macromolecules*, **21**, 3122–4.

6.7. A retarder was added to a radical addition polymerization system at a concentration of 5.1 per cent of the initial concentration of the monomer. The system was thermally initiated by AIBN. The ratio of the rate constant for the reaction of a growing chain with a retarder molecule to the rate constant for propagation is known to be 4.4. It is also known that the product of the reaction with the retarder is unable to react with the monomer. Predict the fractional reduction of the initial polymerization rate produced by the addition of the retarder.

6.8. α-Benzoxylstyrene (BS) acts as a chain transfer agent by an addition-dissociation mechanism represented as

$$-----\cdot + H_2C = C(C_6H_5)OCH_2C_6H_5\,(BS) \;\;\rightarrow -----CH_2 - \dot{C}(C_6H_5)OCH_2C_6H_5$$

$$\downarrow$$

$$\text{Propagation} \leftarrow (C_6H_5)CH_2\cdot \;+\; -----CH_2 - C(=O) - C_6H_5$$

The polymerization of bulk styrene at 333 K was initiated by the thermal decomposition of AIBN, the added concentration of which was $8.6 \times 10^{-3}\,\text{mol dm}^{-3}$.

In the absence of BS, the $\bar{M}_n$ value of the polystyrene produced in the early stages was 1.37×10^5. Given that the value of the chain transfer constant of BS in this system is $C_T = 0.26$, predict the corresponding value of $\bar{M}_n$ of the polystyrene produced in the early stages when the mole fraction of BS is 0.0126. If f (as in eqn (6.4)) = 0.80 for this system, use relevant values of rate parameters from questions above and from Table 6.1 to estimate the total concentration of propagating species in these systems.

Source: Meijs, G. F., and Rizzardo, E. (1988). *Makromolekulare Chemie, Rapid Communications*, **9**, 547–51.

7
Addition polymerization via ionic and coordination mechanisms

Addition polymerizations which proceed via cationic or anionic active centres or by way of coordination to a metal atom are united by the stereospecificity which can lead to high degrees of stereoregularity in the resultant polymer. This contrasts with radical addition polymerizations which tend to yield atactic polymers under industrial synthesis conditions, or sometimes mixed atactic and syndiotactic products when steric interference of bulky groups is significant (as in poly(methyl methacrylate)). An electrically-uncharged radical centre would be expected to have little ability to orientate monomer molecules as they join the chain, particularly as the process takes place at relatively high temperatures when average thermal energies in bonds are large compared to the energy barriers to internal rotation (Section 3.4). Under these circumstances it is only when the substituent X in the monomer $CH_2=CH(X)$ is rather bulky and/or polar that there is any significant influence on the stereoisomeric configuration of the polymer: the syndiotactic form is favoured because it keeps X groups on neighbouring chiral centres as far apart as possible.

The polymerizations concerned in this chapter often proceed at high rates at relatively low temperatures, even well below ambient in the cases of ionic mechanisms. Although relatively few commercial polymers are synthesized via these mechanisms at present, a considerable potential for future usage in producing specialized polymers has been revealed by recent research work. On this basis rather more attention will be paid to the underlying background of these types of systems than might seem to be justified from current industrial significance.

7.1 General features of ionic polymerization

Poly(isobutene) is the principal industrial product from ionic polymerization, cationic polymerization being the route to the high polymer. Radical polymerization is not effective with isobutene $((CH_3)_2C=CH_2)$ because of the efficiency of chain transfer to the monomer; this reflects the relatively high stability of the tertiary butyl radical $((CH_3)_3C\cdot)$ which results from hydrogen atom transfer from the monomer.

The propagating species in cationic and anionic polymerizations are represented as $----\oplus$ and $----\ominus$ respectively. But the charges at the ends of the growing chains cannot exist in isolation since the bulk medium must be electrically neutral. Accordingly a small ion of opposite charge, a *gegenion*, must be located in the vicinity of each propagating chain-end. Thus more proper representations might be $----\oplus X^-$ and $----\ominus M^+$, where X^- and M^+ are the gegenions. Perhaps the most significant consequence of such a pair of charges is its ability as an effective dipole to produce a field which interacts with approaching monomer molecules at relatively long range to orientate them in a particular direction as they join the chain, giving stereoregularity in the resulting polymer.

When polymerization occurs in solution via a radical mechanism, the solvent (assumed to be chemically inert) serves simply as a vehicle for the dissolved species but exerts no major effects on rates or stereochemistry. But in ionic polymerizations the solvent has a very considerable influence and changing it can result in dramatic differences in the nature of the polymer produced. Ionic polymerizations are invariably carried out in solutions in solvents of relatively low polarity, i.e. relatively low relative permittivity (ε). Consider the molecule which is represented as a completely covalent form by RX. In a solvent of high ε, this is considered to tend towards the solvated ions represented as R^+(solv) and X^-(solv), when any particular cation is not associated with any particular anion, i.e. the ions are free. Going from low to high values of ε, intermediate stages of weakening interaction between R^+ and X^- are envisaged. Particular stages may be specified as an *intimate ion pair*, represented as $[R^+X^-]$, and a solvent-separated ion pair, $R^+//X^-$, in which particular ions remain together, in direct contact in $[R^+X^-]$ and with interposed solvent molecules in $R^+//X^-$. When the pair of charges are the end of a propagating chain and its gegenion, the rate at which monomer molecules can add on to the chain will depend upon the ease with which physical contact can be achieved. Evidently more tightly bound pairs and/or bulkier gegenions will impede the interposition of monomer molecules and the rate of their addition (i.e. propagation) will be lower. The overall change in the propagation rate constant, k_p, can be several orders of magnitude for a change of ε by a few units in the typical range concerned of one to 10.

Another contrast with radical addition polymerization is termination; in ionic addition polymerizations this cannot involve the interaction of two growing chain ends, simply because like charges repel one another. Chain transfer with some other molecule takes place in some systems whilst in others termination of a chain takes place in isolation by *divestment*, when the charges (chain end and gegenion) are expelled spontaneously in the form of an ion-paired species leaving the dead polymer.

Ionic polymerization systems are usually considerably more complex than the polymerizations via radical centres. The rapidity of ionic polymerization often creates problems, not the least of which is the extreme sensitivity of rates to the presence of impurities. Rates are sometimes very different in systems which have apparently been made up to be identical under the same conditions. In some instances it is very difficult to specify an overall mechanism with any degree of certainty.

The phenomenon of a '*living polymer*' is a particularly spectacular feature of some ionic polymerization systems, both cationic and anionic instances being known. In this type of system, termination of growing chains simply does not occur provided that impurities, especially oxygen and water, are rigorously excluded. Anionic living polymers have the longer history and have been exploited more than the living cationic polymerizations discovered relatively recently. In a living system, polymerization will cease when the monomer therein has been consumed. The reactor and its contents can be left undisturbed for a considerable length of time without any activity being apparent. But when a new charge of monomer is added, growth of the chains resumes immediately. Significantly the monomer added at this stage need not be the same as that present initially: the resultant ability to synthesize *block copolymers*, such as the *diblock type* $\{A\}_m\{B\}_n$ (referred to as AB block copolymer) or, via a further addition of monomer A, the *triblock type* $\{A\}_x\{B\}_y\{A\}_z$ (referred to as ABA block copolymer), is the main aspect of commercial interest here.

7.2 Cationic polymerization

This section focuses on poly(isobutene), virtually the only polymer synthesized using cationic polymerization on an industrial scale.

Initiation must involve the addition of the monomer to an ionic species bearing a positive charge. Industrially the main initiators are Lewis acids which act as so-called coinitiators (or cocatalysts) with other substances, the initiators (or catalysts). The common combination of this type used in poly(isobutene) synthesis is aluminium chloride and methyl chloride respectively, which achieve charge separation via an overall equilibrium represented as

$$AlCl_3 + CH_3Cl \rightleftharpoons CH_3^+ \cdot AlCl_4^-$$

Another combination used sometimes is boron trifluoride and water, setting up the equilibrium

$$BF_3 + H_2O \rightleftharpoons H^+ \cdot BF_3OH^-$$

When the final product is to be butyl rubber, a few per cent of isoprene $(CH_2=C(CH_3)-CH=CH_2)$ or other diene is added with isobutene. The resultant copolymer then incorporates enough double bonds in the isoprene residues $(-CH_2-C(CH_3)=CH-CH_2)$ to allow it to be vulcanized

(Fig. 3.4). A typical synthesis on an industrial scale involves feeding the required mixture of isobutene and diene dissolved in methyl chloride into the stirred reactor which is cooled to around 178 K using externally circulating liquid ethylene. Initiation is achieved by the addition at the prescribed flowrate of a solution created by dissolving anhydrous aluminium chloride in methyl chloride. The polymer emerges in a slurry which is removed continuously from the reactor to be pumped into a large tank containing hot water which is vigorously agitated. Volatile contaminants, such as unreacted monomer and methyl chloride, are recovered from the vapour which is extracted from the tank above the water level. Aluminium compounds dissolve into the water. Ultimately the raw polymer is isolated by filtering it from the water. Washing, drying, and processing produces the finished butyl rubber subsequently.

The propagating chain is both a carbocation and a *carbenium ion*, but it is only when this is of a tertiary nature that it has sufficient thermodynamic stability to promote cationic addition polymerization. Thus

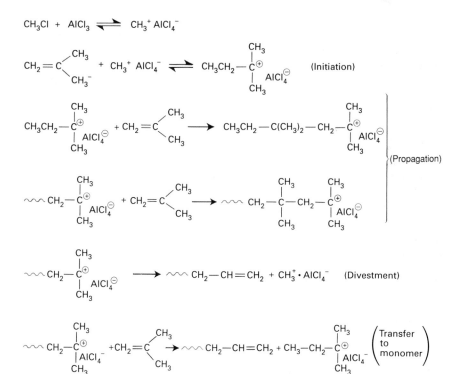

Fig. 7.1. Representation of the steps involved in the mechanism of the cationic addition polymerization of isobutene initiated by the aluminium trichloride–methyl chloride system.

it is only isobutene amongst the common alkenes which can be polymerized in this way. A straightforward mechanism which indicates the likely elementary steps involved in the cationic polymerization of isobutene using the initiation system discussed above is shown in Fig. 7.1.

The $\bar{M}_n$ of the polymer for butyl rubber production is typically of the order of 3×10^5. Following vulcanization its main usage is in the manufacture of tyres for vehicles. Butyl rubber has two main advantages in comparison with natural rubber: it remains flexible to much lower temperatures (220 K or so) and it is more resistant to attack by components of the air, principally ozone, i.e. it does not 'perish' so quickly.

A general kinetic scheme for cationic polymerization can be written in terms of C (the catalyst or initiating system), M (monomer), $----\oplus$ (C^-) (a growing chain and gegenion) and Tr (some transfer agent). This predicts the behaviour of many but by no means all systems.

$$
\begin{array}{ll}
\text{M} \quad\quad + \text{ C} \xrightarrow{k_i} \text{M}^+(\text{C}^-) & \text{(initiation)} \\
\text{M}^+(\text{C}^-) + \text{M} \xrightarrow{k_p} \text{M}_2^+(\text{C}^-) & \\
\quad\quad \cdots\cdots\cdots\cdots\cdots\cdots\cdots\cdots\cdots\cdots & \text{(propagation)} \\
----\oplus (\text{C}^-) + \text{M} \xrightarrow{k_p} ----\text{M}^+(\text{C}^-) & \\
----\oplus (\text{C}^-) + \text{Tr} \xrightarrow{k_{tr}} ----+\text{Tr}^+(\text{C}^-) & \text{(termination} \\
& \quad \text{by transfer).}
\end{array}
$$

The corresponding rate constant is indicated above the reaction arrow.

A steady state concentration of propagating chains is expected so that the rates of initiation and (r_i) and termination are equated

$$ r_i = k_i[\text{M}][\text{C}] = k_{tr}[\text{Tr}] \cdot \Sigma[----\oplus(\text{C}^-)] $$

which on rearrangement yields the expression for the total concentration of all growing chains

$$ \Sigma[----\oplus(\text{C}^-)] = (k_i/k_{tr}) \cdot [\text{M}][\text{C}]/[\text{Tr}]. $$

The overall rate of polymerization is identical with the total rate of propagation and is thus defined as

$$ \text{rate} = k_p[\text{M}] \cdot \Sigma[----\oplus(\text{C}^-)] = (k_p \cdot k_i/k_{tr}) \cdot [\text{M}]^2[\text{C}]/[\text{Tr}] $$

with substitution of the equation immediately above. The average degree of polymerization ($\overline{\text{DP}}$) is expressed as the ratio of the rates of propagation and terminations (transfer here) as

$$ \overline{\text{DP}} = \{k_p[\text{M}] \cdot \Sigma[----\oplus(\text{C}^-)]\}/\{k_{tr}[\text{Tr}] \cdot \Sigma[----\oplus(\text{C}^-)]\} $$

$$ = (k_p/k_{tr}) \cdot [\text{M}]/[\text{Tr}]. $$

Thus the value of $\bar{M}_n$ ($\propto \overline{\text{DP}}$) achieved in cationic polymerization is independent of the rate of initiation and thus also of the concentration of the initiator. This contrasts sharply with radical addition polymerization (see eqn (6.3)).

Table 7.1 Arrhenius parameters of cationic polymerizations of alkyl vinyl ethers initiated by triphenylmethyl salt

R	E_i^*	E_p^*	E_{tr}^*	Predicted E_{obs}^*
CH_3	36	46	63	19
C_2H_5	28	45	54	19
$i-C_3H_7$	23	15	24	14
$i-C_4H_9$	39	23	41	21

*kJ mol^{-1}.
Added concentration ranges: alkyl vinyl ether $(CH_2=CH(OR))$, $(2-15) \times 10^{-2}$ mol dm^{-3} triphenylmethyl salt, $(5.5-8.0) \times 10^{-5}$ mol dm^{-3}.
Based upon data given by Subira, F., Vairon, J. P. and Sigwalt, P. (1988). *Macromolecules*, **21**, 2339–2346.

The predicted form of the rate law for cationic polymerization when transfer to an added agent dominates the termination of chains is

$$-d[M]/dt = k_{obs}[C][M]^2/[Tr].$$

The observed rate constant, k_{obs}, is identified with $(k_p k_i/k_{tr})$, from the analysis above. In the corresponding Arrhenius equation, $k_{obs} = A_{obs}$ $\exp(-E_{obs}/RT)$, the observed activation energy, E_{obs}, is therefore predicted to be expressed in terms of the activation energies for the elementary steps of initiation (E_i), propagation (E_p), and transfer (E_{tr}) as

$$E_{obs} = E_p + E_i - E_{tr}. \tag{7.1}$$

Table 7.1 shows representative values of these activation energies for the cationic polymerization of a series of alkyl vinyl ethers $(CH_2=CH(O-R))$, when a triphenylmethyl salt, $(C_6H_5)_3C^+SbCl_6^-$ is the initiating agent in dichloromethane solutions at temperatures between 233 K and 273 K. In this case transfer to monomer was the predominant termination mode. The key point for the low temperatures at which the polymerizations may proceed is that the values of E_{obs} are rather small in comparison with those for radical addition polymerization (Table 6.1). Relatively low values of E_i and relatively high values of E_{tr} are apparently mainly responsible for these low values of E_{obs}.

7.3 Anionic polymerization

Aside from living polymers, anionic polymerization is of only minor significance in the industrial context. The main feature of interest of some other polymers synthesized in the laboratory by anionic mechanisms is the stereoregularity: an example of this is poly(methylmethacrylate), as discussed in connection with Figs. 2.7–2.9.

The general features of many anionic polymerization systems are similar to those of many cationic polymerization systems. Now the chain is

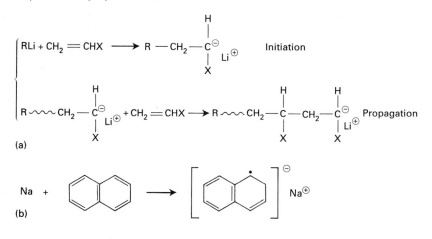

Fig. 7.2. Representation of (a) initiation and propagation steps in the cationic addition polymerization of the vinyl monomer $CH_2=CH(X)$ by an alkyl lithium (RLi) and (b) the formation of the initiating ion-pair in the sodium/naphthalene system.

negatively charged and the gegenion is a cation, such as Li^+. The commonest initiators are simple organometallic compounds such as butyllithiums, which are easily soluble in the usual organic solvents concerned, such as toluene and tetrahydrofuran. In general the initiation step involves the addition of the organometallic molecule across the $C=C$ bond of the monomer, which is the point at which true charge separation occurs. When the organolithium compound is denoted by RLi, initiation and the subsequent propagation step with the vinyl monomer $CH_2=CH(X)$ may be shown as in Fig. 7.2(a).

Another initiating system of interest is created when sodium metal is dissolved into a solution of naphthalene in tetrahydrofuran. The result is the generation of the naphthalene anion radical, as shown in Fig. 7.2(b), which gives the solution its characteristic green colour.

Polyoxymethylene (also known as acetal resin) is a polymer of formaldehyde ($H_2C=O$) and such carbonyl compounds are not susceptible to polymerization by radical addition mechanisms. However polyoxymethylene can be synthesized through ionic routes. The commercial product comes commonly from a solution of formaldehyde in a hydrocarbon solvent, such as n-hexane, at temperatures extending down to 220 K. A variety of initiators are used, including metal alkoxides and metal amides, to create the anionic growth centres. As it forms, the polymer precipitates out and is removed continuously as a slurry from the reactor. It is usually necessary to stabilize the chains by a process known as 'end-capping'. As synthesized without end-capping each chain is terminated by a hydroxyl ($-OH$) group, which originates in transfer of a hydrogen atom from another molecule to the $-CH_2-O^-$ active end. Acetic anhydride is added, which results in the replacement of these hydroxyl groups by acetyl groups ($-O-C(=O)CH_3$) and the hydrogen atom of the hydroxyl group is removed in the form of acetic acid. This end-capping with acetyl groups prevents the depolymerization of polyoxymethylene which would otherwise commence from the hydroxyl groups when the polymer was exposed to temperatures approaching its rather low ceiling temperature (see Fig. 1.11). End-capped polyoxymethylene is a thermoplastic with a relatively high degree of crystallinity (around 75 per cent is typical) and also a relatively high value of T_m (453 K). The good properties of the bulk polymer are the possession of both high impact and tensile strengths and the abilities to retain both shape and finish. In fact end-capped polyoxymethylene has significant usage as an engineering plastic, substituting for metals like brass and cast iron used in former times.

Undoubtedly the most interesting systems of this type, which have achieved some commercial success already and offer great potential, are living anionic polymerizations. These have been used to synthesize block copolymers, particularly those ABA triblock copolymers which have the properties of *thermoplastic elastomers* (see Section 8.4). Of these, SBS (styrene–butadiene-styrene) triblock copolymer is the major commercial product at present and is thus used for exemplification.

The addition of butadiene to the sodium–naphthalene initiator system (Fig. 7.2(b)) in tetrahydrofuran at temperatures below ambient results in the generation of a living dianion, with charges and thus active propagation sites at both ends of the chain, as shown in Fig. 7.3. What is represented is that the transfer of an electron from the naphthalene radical anion to a butadiene molecule gives rise to the radical anion form of the latter. This species dimerizes via coupling of the unpaired electrons to give the dianion which is able to grow from both ends simultaneously through addition of butadiene residues. The desired blocks of polybutadiene contain 1000–1500

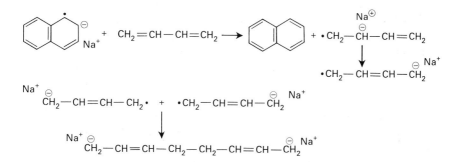

Fig. 7.3. Representation of the mechanism of generation of a dianion in the system of sodium/naphthalene and butadiene.

monomer residues when propagation is stopped by cutting off the supply of butadiene to the reactor. Then styrene is added and polymerization restarts and is allowed to continue until 100–150 styrene residues form the end blocks of each chain. Finally the chains are 'killed' by addition of a terminating agent, sometimes water.

It is of interest to point out that the polymer produced in a living system has a low dispersity, i.e. a narrow distribution of molecular masses about the mean. When all chains have been initiated at virtually the same time and each has an equal probability of propagating at any instant and no probability of termination or chain transfer, the low polydispersity is hardly surprising. In fact anionic polymerization via living systems is usually the source of the 'standard samples' of almost single molecular mass which are used in the calibration of gel permeation chromatography systems (Section 2.2.2).

7.4   Coordination polymerization

7.4.1   General features

The most important systems in this connection are *heterogeneous* (i.e. involve the crossing of phase boundaries) and comprise a catalytically active solid surface working in conjunction with a liquid or a gaseous phase. The products are unique, in having strictly unbranched (i.e. linear) chains which are often stereoregular when there are chiral centres in the chain backbone.

The first coordination catalysts emerged around 40 years ago, when Ziegler discovered combinations of transition metal compounds and organometallic compounds which were able to catalyse the polymerization of ethylene at relatively low temperatures and pressures. Natta then found that these types of catalysts polymerized higher alkenes to yield stereoregular products. These so-called Ziegler–Natta catalysts are now in commercial usage to the extent that isotactic polypropylene has become a *commodity plastic*, that is, a plastic able to be produced in high volume at low cost and accordingly used in such common forms as rope or sacking. In fact, it is only the isotactic form of polypropylene which is available in general (Fig. 1.5).

The other coordination polymerization system which has achieved major significance is used in what is known as the Phillips process. Here the active catalyst is composed of isolated chromium–oxygen bridges $(-O-Cr-O-)$ anchored to the surface of a support which is usually silica (SiO_2). Large proportions of what are known as high-density polyethylene (HDPE) and linear low-density polyethylene (LLDPE) (Fig. 1.5) are produced in this type of system. HDPE and LLDPE are distinguished from the low-density polyethylene (LDPE) which is produced by radical addition polymerization (Section 6.6). The differences between the backbone

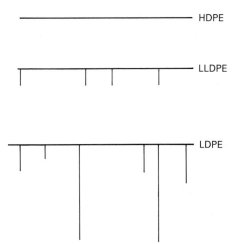

Fig. 7.4. Simple models of the backbone structures of the forms of polyethylene known as HDPE, LLDPE, and LDPE. The main chain is represented as a horizontal line, whilst the location and lengths of the branches are indicated by the vertical lines.

structures of these forms of polyethylene are represented in simple terms in Fig. 7.4.

HDPE has no branches, LLDPE has short and uniform branches whilst LDPE has branches of more random length, some of which are short (produced by backbiting (Fig. 6.7)) and others may even be polymeric (produced by chain transfer during synthesis).

In both the Ziegler–Natta and Phillips processes, it is generally accepted that polymer chains are synthesized whilst chemically bonded to localized sites of very specific structure. There are single atoms of titanium and chromium respectively at each active site and these are in oxidation states, generally believed to be III, which represent coordinative unsaturation, i.e. there are vacant 'd' orbitals which can accept rather easily pairs of electrons donated by other molecules to form coordination bonds. There are two such vacant orbitals at each site initially, allowing a polymer chain and a monomer molecule to be bound there simultaneously. Figure 7.5 indicates the main features of the propagation act in general terms, using □ to represent a coordinative vacancy for a pair of electrons.

According to this set of diagrams, a growing chain (represented by a wiggling line) is coordinatively bound to a metal atom which has another coordinative vacancy (Fig. 7.5(a)); this is able to interact with the π-electron density of a monomer (ethylene here) molecule (Fig. 7.5(c)). This π-electron density is then pulled out of the C=C bond to start to create bonds between one of the carbon atoms of the monomer and the metal atom and between the other carbon atom of the monomer and the innermost carbon atom of

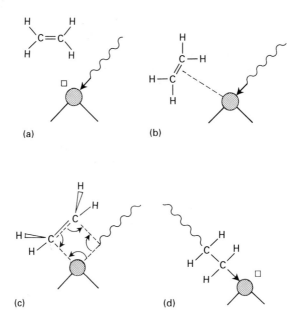

Fig. 7.5. Representation of the general mechanism of propagation in coordination polymerization of ethylene. The filled circle represents the metal atom and □ indicates the vacancy for two electrons. The preexisting polymer chain is shown by the wiggly line and its coordinative bond to the metal atom is indicated by an arrow. Dashed lines represent partial bonds and curved arrows indicate the directions of movement of electron density.

the existing chain. Electron density from the original coordination bond which anchors the polymer chain also moves to strengthen these new bonds (Fig. 7.5(c)). Continuation of this redistribution of electron density achieves the effective insertion of a new monomer residue between the pre-existing polymer chain and the metal atom (Fig. 7.5(d). It is then obvious that the resultant chain from many such acts must be linear. There is no mechanism for branching to occur since the chain grows effectively from the inside, with what may be identified as the active site being the carbon atom which is directly bonded to the metal atom.

Now that the common features of these systems have been outlined, it is appropriate to develop their more specific aspects in the following subsections.

7.4.2 Ziegler–Natta and related processes

The commercial production of isotactic polypropylene uses an insoluble catalyst, titanium chloride ($TiCl_4$), supported on magnesium chloride

($MgCl_2$); a typical loading is 0.5 moles of titanium per kg of solid. Now homogeneous Ziegler–Natta systems with the catalysts in solution are well known and the mechanisms of polymerization and the natures of the active sites on these are considered as quite well understood. The main recourse in attempting to specify corresponding aspects for the heterogeneous systems of main industrial interest is to presume that there are valid analogies between the two types of system.

There are two main components in an industrial Ziegler–Natta polymerization system. The first is the solid catalyst, the second is a combination of a *cocatalyst*, typically triethylaluminium (($C_2H_5)_3Al$), and an electron donor such as diphenylmethoxysilane (($C_6H_5)_2Si(OCH_3)_2$), the latter referred to as the *external modifier*. The active site on the catalyst surface is believed to centre on a titanium atom, which has not achieved the state with six bonds directed octahedrally from it which would correspond to its full coordination. What it has is four bonds to chlorine atoms, two of these also being bonded to a magnesium atom in the solid surface. Prior to the addition of the cocatalyst, another chlorine atom would have been relatively weakly bonded to this titanium atom. The activation of this site for polymerization involves the reaction of the cocatalyst there: in the case of triethylaluminium, the reaction substitutes an ethyl group for this 'loose' chlorine atom, which is removed as ($C_2H_5)_2AlCl$. The sixth potential bond, needed to produce full coordination of the titanium atom, is not established, so that a coordinative vacancy exists. The ethyl group at this stage may be regarded as the first part, which will eventually become the terminal monomer residue, of the polymer that will grow. The first propagation act can be envisaged to involve the arrival of a monomer molecule at this coordinative vacancy. Then the ethyl group transfers across to form a new $C-C$ bond with what becomes the outer of the two carbon atoms of the residue of the monomer molecule. Further acts of propagation then follow the pattern indicated in Fig. 7.5. It is believed that the external modifier functions by changing the action of other sites which, although they are active for polymerization, do not give rise to the isotactic polymer in the case of propylene. It is widely accepted that the molecules of the external modifier attach themselves to these sites by coordinative bonds and may even convert them into sites at which the isotactic polymer is synthesized. This seems to be possible because the external modifiers used in practice result in polypropylene with ≥ 95 per cent of isotactic nature being produced at much the same rate at which overall polymerization occurs in their absence.

The chains which would grow in Ziegler–Natta systems would usually be excessively long if a chain transfer agent was not added. That in commonest use is simply molecular hydrogen (H_2), which is believed to donate a hydrogen atom to terminate and detach the chain (now dead) from the titanium atom, leaving the other hydrogen atom in its place effectively.

The industrial scale production of polypropylene is usually carried out with the catalyst present in a slurry containing about 10 per cent solids in a liquid alkane (say heptane). Batch or continuous (flowing) modes of operation are possible. In the typical batch system, the central feature is a reactor which can be pressurized, to which is added the appropriate amounts of catalyst slurry, additional liquid alkane, cocatalyst, and external modifier. Then propylene is introduced to maintain a moderate pressure, two to six times atmospheric, with hydrogen ($\sim$0.1 atmospheric pressure) as polymerization is proceeding at $\sim$335 K for some 5 to 8 h. During this time the slurry in the reactor is well agitated. The reactor temperature is well below T_m ($\sim$443 K) of isotactic polypropylene, which appears as a solid suspension therefore. The product is removed from the reactor and centrifuged to remove any liquid alkane and the small amount of atactic polypropylene which is also formed, but is dissolved in the alkane in contrast to the insolubility of the isotactic main product. Residual catalyst material is removed by treatment with methanol containing a small concentration of hydrogen chloride. Finally washing with pure methanol gives raw isotactic polypropylene, ready for blending with the additives which improve the stability of the product supplied commercially.

It is worthwhile mentioning here that various organometallic complexes, particularly of zirconium and hafnium, have been used as catalysts for the polymerization of propylene in laboratory studies. Some of the polypropylenes had an isotactic content which exceeded 90 per cent, but perhaps more interesting were others which were almost totally syndiotactic or atactic. For example, one synthesis of almost pure syndiotactic polypropylene proceeded as follows. The catalyst was isopropyl(cyclopentadienyl-1-fluorenyl) zirconium in conjunction with methylaluminoxane and both were dissolved in toluene. Methylaluminoxane is the reaction product of trimethylaluminium with the appropriate amount of water and has the general formula $CH_3(-Al(CH_3)-O)_n-CH_3$. When the solution was held under propylene at 293 K for 2 h in an autoclave reactor, the product was polypropylene in which the syndiotactic pentad fraction (rrrr, see Fig. 1.10) was 0.915 as measured by ^{13}C NMR (Section 4.2). On the other hand, when the catalyst was biscyclopentadienyl zirconium dichloride in association with methylaluminoxane, the product was almost pure atactic polypropylene.

7.4.3 Phillips processes

Linear low-density polyethylene (LLDPE) and high-density polyethylene (HDPE) are the main products obtained via Phillips catalytic systems. LLDPE is a copolymer of ethylene predominantly with residues of another small 1-alkene. As shown in Fig. 7.4 the resultant polymer chain has a few short branches, all of the same length since they are simply the rest of the

1-alkene molecule if the C=C bond is discounted. For example with one molecule of 1-pentene per $(m + n)$ molecules of ethylene, the average result of the LLDPE synthesis process can be represented as

$$(m + n)CH_2{=}CH_2 + CH_3CH_2CH_2CH{=}CH_2 \rightarrow$$

$$(-CH_2-CH_2)_m-CH(CH_2CH_2CH_3)-CH_2-(CH_2-CH_2-)_n.$$

A catalyst for the Phillips process may be produced by a procedure which commences with the impregnation of silica gel with an aqueous chromate solution to give a chromium loading of ~1 per cent after drying. Following heating to $\geq 800\,K$ in air and cooling, the chromium is present in a dispersed form bonded to the surface of the silica gel: Fig. 7.6 is believed to represent a typical structure at this stage. The creation of the active site for polymerization involves the reduction of the Cr(VI) in this structure to Cr(III) in the structure on the right of Fig. 7.6. This reduction can be effected via the partial oxidation of ethylene to formaldehyde as indicated. The surface structure on the right, indicated as having two coordinative vacancies (represented by $\square$) for pairs of electrons, is believed to induce the processes represented in general terms in Fig. 7.5.

An industrial-scale Phillips process works as follows typically. The catalyst is present in a slurry in an alkane (such as cyclohexane) in a flowing reactor. Ethylene is pumped into the base of the reactor at around 35 times atmospheric pressure and dissolves into the liquid phase in which it polymerizes at temperatures of 400–440 K. The hot liquid leaving the reactor contains the polymer in solution. It is first filtered to remove suspended catalyst particles and then it is cooled to precipitate out the polymer. In a representative sample of HDPE from a Phillips process, some chains have relative molecular masses as low as 10^3 whilst others may be as high as 10^6. The polydispersity of this typical product is evidently relatively high, but it can be varied to some extent by altering the conditions during the synthesis.

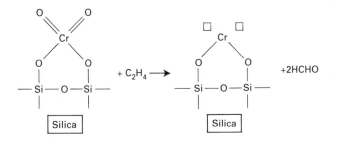

Fig. 7.6. Representation of the reduction of a Cr(VI) site (*left*) to a Cr(III) site by ethylene.

Reaction rates in Phillips systems are usually considerably lower than those in the corresponding Ziegler–Natta system. Another difference is that when higher alkenes are polymerized in Phillips systems, the products do not have high degrees of stereoregularity. These are the principal reasons which restrict the industrial exploitation of the Phillips process to ethylene as the monomer. Moreover Phillips processes cannot be worked in the batch mode, only continuously.

Exercises

7.1. The cationic polymerizations of alkyl vinyl ethers referred to in Table 7.1 follow rate laws which are first order in both the initiating triphenylmethyl salt and the monomer. How may this be explained?

7.2. The cationic polymerization of styrene in solution in nitrobenzene at 298 K has been initiated by the tin tetrachloride/tertiary-butyl chloride system which may be considered to generate the cationic centre via

$$t\!-\!C_4H_9Cl + SnCl_4 \rightleftharpoons t\!-\!C_4H_9^+ \cdot SnCl_5^-.$$

When both $t\!-\!C_4H_9Cl$ and $SnCl_4$ were added to give a concentration of each of $1.00 \times 10^{-3} \, mol \, dm^{-3}$ and styrene was present at a concentration of $0.800 \, mol \, dm^{-3}$, the rate of polymerization was measured as $1.28 \times 10^{-3} \, mol \, dm^{-3} s^{-1}$.

(a) What fraction of the added tin is involved in the active propagating centres?

(b) What is the value of the standard equilibrium constant for the initiating system if the rate constant for the propagation step (obtained from other work) is $200 \, dm^3 \, mol^{-1} s^{-1}$?

Source: Plesch, P. H. and Shamlian, S. H. (1990). *European Polymer Journal*, **10**, 1113–20.

7.3. An anionic polymerization system is composed of $1.000 \, g$ of methyl-methacrylate (monomer) and $0.01281 \, g$ of tertiary-butyl lithium (initiator) dissolved in $0.010 \, dm^3$ of toluene. The polymerization was carried out at 195 K and the yield of poly(methylmethacrylate) was 89 per cent by weight of the initially added monomer. On the assumption that every molecule of t-butyl lithium produces an active propagation centre, estimate $\bar{M}_n$ of the polymer if this is a living system.

Source: Kitayama, T., Shinozaki, T., Masuda, E., Yamamoto, M. and Hatada, K. (1988). *Polymer Bulletin*, **20**, 505–10.

7.4. Tetrahydrofuran can be polymerized by an ionic mechanism but not by a radical polymerization system with thermal initiation. Explain this by reference to Fig. 1.11.

7.5. An ABA triblock copolymer, with monomer A being butadiene ($CH_2=CH-CH=CH_2$) and monomer B being isoprene ($CH_2=C(CH_3)-CH=CH_2$), has been synthesized by an anionic living mechanism in benzene solution at 298 K, with sequential addition of monomers. The initiator was secondary-butyl lithium. The polymer was 'killed' by addition of tertiary butanol. The $\bar{M}_n$ of the ABA copolymer was 1.57×10^5 when its weight fraction of polybutadiene was 0.27.

(a) Explain how tertiary butanol is likely to 'kill' chain growth.

(b) Deduce the structure of the average chain in terms of the number of monomer residues in the blocks if it is assumed to be symmetrical about the centre of the chain.

Source: Seguéla, R. and Prud'homme, J. (1989). *Polymer*, **30**, 1446–55.

7.6. Draw a set of diagrams corresponding to Fig. 7.5 to illustrate the action of a heterogeneous catalyst produced from titanium chloride and magnesium chloride in the Ziegler–Natta polymerization of propylene.

7.7. Figure 7.7 shows circles and an irregular shape which represent the relative contact sizes of a chlorine atom, a titanium(III) atom, and the propylene molecule (molecular orientation indicated within) respectively. Trace these through onto separate paper and cut out shapes for *six* Cl, *two* Ti atoms, and *two* propylene molecules. Arrange two pairs of three, one and one of these respectively to touch in the appropriate way to represent a cross section of the complex formed through interaction of the π-electrons of the C=C double bond with $TiCl_3$ effectively. Show that there is only one orientation of the propylene molecule which corresponds to a comfortable fit. What is the significance of this?

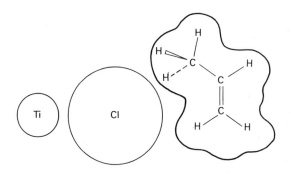

Fig. 7.7. Relative contact sizes of titanium (III) (Ti) and chlorine (Cl) atoms, and of a propylene molecule (in the plane of the C=C bond).

7.8. Syndiotactic polymerization of styrene is induced by methylaluminoxane (MAO) and titanium tetrabutoxide (TTB) in toluene as the solvent. In the system at 323 K made up with $0.0250 \, dm^3$ of toluene, $0.0150 \, dm^3$ of styrene, 4.0×10^{-3} moles of MAO, and 4.5×10^{-5} moles of TTB, the following results were obtained.

Time (min)	7	25	45	60
Yield of polymer (mg)	17	60	107	143
$10^{-3} \bar{M}_n$ of polymer	11	12	13	12

What may be deduced?

Source: Oliva, L., Pellecchia, C., Cinquina, P. and Zambelli, A. (1989). *Macromolecules*, **22**, 1642–5.

8
Properties of common polymers in bulk

The means through which the lengths and architectures of polymer chains may be tailored have been discussed in the preceding chapters. Attention is turned now to the effects on the bulk properties that result from variations of the chain structures. Also of interest are particular modifications of polymers in bulk which produce dramatic changes in their natures, enhancing ranges of applications.

8.1 Crystallinity

This section may be considered to take up where Section 3.5 left off.

The maximum degree of crystallinity of a thermoplastic in bulk which can be achieved at a specified temperature above depends T_g depends on the lengths of the chains and hence on the average relative molecular mass. This is illustrated in Fig. 8.1 (on this scale $M = \bar{M}_n \approx \bar{M}_w$).

The downward trend of the maximum degree of crystallinity as the chains become longer may be explained in terms of increasing entanglement. Thus the disentanglement required prior to alignment of chain segments in crystalline regions becomes increasingly difficult with lengthening chains. Moreover microscopic features such as 'loops' and 'knots' are points of resistance which become more frequent the longer the chains. Thus extrication of even segments of a chain and alignment with other extricated segments in a crystallite during the time in which the thermoplastic is annealed close to T_m becomes less and less likely at higher values of M.

In keeping with the increasing proportion of amorphous material implied by Fig. 8.1, the macroscopic density of bulk polyethylene decreases as the chains become longer, typically from 0.99 to 0.92 kg dm^{-3} as $\bar{M}_n$ goes from 10^4 to 10^8. Table 8.1 lists densities of some common polymers in fully amorphous and fully crystalline forms, the latter of microscopic significance only.

What is known as very low-density polyethylene (VLDPE) has been made by copolymerizing several 1-alkenes with ethylene in a Phillips system (Section 7.4.3) to give more short-chain branching than in LLDPE (Fig. 7.4) and with side chains of variable length. These features discourage

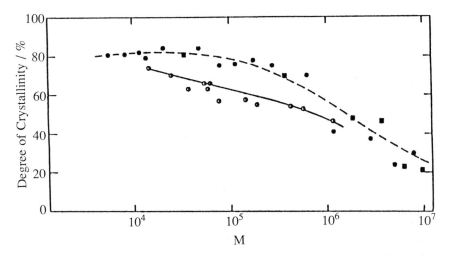

Fig. 8.1. Plot of the degree of crystallinity as a function of relative molecular mass (*M*) under isothermal crystallization conditions. ● is linear polyethylene ■ is poly(ethylene oxide), and half-filled circles are poly(tetramethyl-p-silphenylenesiloxane). Reprinted with permission from Mandelkern, L. (1990). The structure of crystalline polymers. *Accounts of Chemical Research*, **23**, 380–6. Copyright (1990) American Chemical Society.

Table 8.1 Densities of polymers in bulk and in crystalline and amorphous regions

Polymer	Density (kg dm^{-3})		
	Typical bulk	**Crystalline**	**Amorphous**
Polyethylene (LDPE)	0.91–0.93		
Polyethylene (LLDPE)	0.93–0.94	1.00	0.855
Polyethylene (HDPE)	0.95–0.96		
Polyethylene (VLDPE)	0.90–0.905		
Poly(ethylene terephthalate)	1.41	1.455	1.335
Poly(tetrafluoroethylene)	2.19	2.302	2.000

crystallinity so that VLDPE is more amorphous than even LDPE: its density is close to that of polypropylene (typical bulk density of 0.90 kg dm^{-3}), the least dense of all bulk polymers in large-scale usage.

High or low proportions of stereoregularity in the chains would be expected to govern the degree of crystallinity which can be achieved in the bulk polymer. Atactic chains, with random orientations of substituent groups along the chains, are least likely to be crystalline, simply because of the difficulty of packing such irregular chains closely together in a lattice. Atactic polymers are almost wholly amorphous in the bulk accordingly, for example the forms of polystyrene and polyacrylonitrile in common usage. Conversely stereoregular forms, particularly the isotactic chains resulting

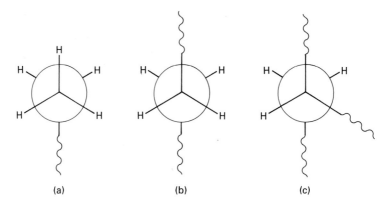

Fig. 8.2. Newman projections of *trans* conformations of a C—C bond of polyethylene at (a) the end of a chain, (b) within a linear chain, and (c) at a branching junction. The wiggly lines represent polymer chains.

from coordination polymerizations (Section 7.4), can be highly crystalline. Table 8.2 summarizes the main differences in the properties shown by bulk samples of common thermoplastics when these are composed predominantly by isotactic or atactic chains.

Discussion of the origins of some of these properties appears in Section 3.3 and others can be understood in terms of the crystalline regions being more resistant to disruption than amorphous regions. The resistance of isotactic material to dissolution in contrast to the ready solubility of the atactic polypropylene in methanol has been mentioned in Section 7.4.2. This is explained in terms of the tangled chains of the atactic form having relatively low cohesive energy and being more easily penetrated by the solvent.

Other relevant aspects of crystallinity in polymers have been discussed in Sections 3.1, 3.2, 3.5, and 4.3.

8.2 Variation of glass transition temperature (T_g)

The influence of the structure of the constitutional repeating unit (CRU) on T_g is discussed in Sections 3.3, 3.4 and 3.6. The interest here is in how the lengths of the chains and the structure of the bulk polymer can affect T_g.

The glass transition occurs in the amorphous regions of a thermoplastic and throughout an elastomer. Below T_g the entire material is hard and glassy. When a thermoplastic is above T_g (and below T_m) the bulk solid is flexible due to the ease of segmental rotations in the amorphous regions surrounding crystallites in which the chains are rigid and immobile. When

Table 8.2 Main differences in bulk properties of the same polymer in predominantly isotactic or atactic forms

Isotactic	Atactic
Higher density	Less brittle
Higher softening temperature	Lower impact strength
Translucent/opaque	Optically transparent
More rigid	
Less soluble in organic solvents	
Fibre-producing potentials	

an elastomer is above T_g, it is rubbery because almost all segments can rotate easily in the virtual absence of crystallites.

It is the relative mobilities in terms of internal rotation at the ends of chains and at $C-C$ bonds within the backbone which are important in connection with the variation of T_g with average relative molecular masses. It is also necessary to consider the hindrance to internal rotation in adjacent $C-C$ bonds imposed at points where the chain backbone branches. Simple Newman projections of the trans conformations (Fig. 3.5) for a $C-C$ bond in these parts of a polyethylene chain are shown in Fig. 8.2.

It is evident that there is least interference to internal rotation about the backbone $C-C$ bond concerned at the end of the chain (Fig. 8.2(a)), where there is only one polymer chain substituent. Moreover the terminal methyl group can be expected to indulge in fairly violent internal rotations, when it is not 'anchored' by any polymeric substituent, as is the case with a $C-C$ bond within the chain (Fig. 8.2(b)). Accordingly the 'thrashing about' of the end group sweeps out a relatively large void volume in its vicinity from which parts of other chains are excluded effectively. Rotations of segments within chains also creates void volume in the bulk, but because the individual $C-C$ bonds are relatively hindered by the two polymeric substituents they have lower void volumes associated with them than those at the ends of the chain.

Consider at first the hypothetical situation in which the whole of an amorphous sample of a polymer was composed of just one enormously long chain. There would be only two chain ends and thus the void volume within the sample would be the minimum. Voids separate chain segments from one another and are essential for segmental rotation: there must be an adjacent void volume for a chain segment to rotate into, in turn leaving a void for another segment and so on. It is then clear that the greater is the void volume, the easier is segmental rotation overall and thus the higher is likely to be the fraction of all segments which must be unable to rotate for this to cause the entire assembly to seize up (Section 3.4). In turn this implies a lower total content of energy required to keep the critical fraction of chain segments mobile and thus a lower value of T_g. Returning to the enormously long chain, imagine that there is a miniature device which can enter this and cut one of the C—C bonds in the backbone. The sample now contains four chain ends and it expands a little because of their relatively large void volumes. The increased void volume is expected to make segmental rotation easier in the sample as a whole. In turn these two chains can be cut to make segmental rotation easy in more and more places as the number of chain ends and hence total void volume increases again. Equally each cut lowers $\bar{M}_n$. This very simplified approach allows the prediction to be made that the value of T_g should fall as $\bar{M}_n$ falls, which simply reflects the expectation that segmental rotation becomes easier as the chains become shorter. In several instances the linear relationship expressed by eqn (8.1) has been found to hold in good approximation

$$T_g = T_g^\infty - C/\bar{M}_n \qquad (8.1)$$

T_g^∞ and C are constants for a particular polymer when the degree of branching is constant. Table 8.3 lists some representative values.

Table 8.3 Values of parameters for eqn (8.1) for some amorphous (atactic) polymers

Polymer	T_g^∞ (K)	C(K)
Polystyrene	373	1.74×10^5
Poly(α-methylstyrene)	442	3.26×10^6
Poly(methylmethacrylate)	388	2.21×10^6

The linearity of plots of T_g versus $(\bar{M}_n)^{-1}$ is good provided that $\bar{M}_n > 5 \times 10^4$ or so in general. The value of $C/\bar{M}_n$ at this lower limit is 35–65 K for the entries in Table 8.3, so that T_g can be significantly lower for a sample of polymer composed of shorter chains than for a sample of the same polymer composed of much longer chains.

A very important phenomenon is the lowering of T_g which can occur when a nonpolymeric liquid is incorporated into some polymers. This is known as *plasticization* and its main usage is with poly(vinyl chloride). The liquid put into the polymer (i.e. the *plasticizer*) must be involatile in this case and dioctyl phthalate (DOP) (di-2-ethylhexyl phthalate) is usually used. The spectacular effect on T_g is evident from the data in Table 8.4.

Table 8.4 Variation of T_g of poly(vinyl chloride) with content by weight of dioctyl phthalate (DOP) plasticizer

% DOP w/w	0	10	40	50
T_g (K)	354	333	267	244

Pure poly(vinyl chloride) is very suitable for making articles of a fixed shape, such as pipes and guttering, since it is a rigid material with T_g well

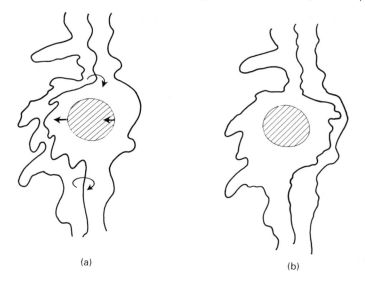

(a)   (b)

Fig. 8.3. Diagram explaining plasticization by a small volume of involatile liquid (hatched blob) interposed between polymer chain segments (lines).

above ambient temperatures. However plasticized poly(vinyl chloride) containing 40–50 per cent by weight of DOP is a flexible material with T_g well below ambient; it is thus suitable for making waterproof clothing.

Plasticization may be explained by reference to the simple model shown in Fig. 8.3. The barred and roughly circular area represents a 'blob' of the plasticizer liquid, composed by a large number of molecules which are each very small compared to those of the polymer. Sections of three polymer chains are represented in the diagram by irregular lines. For simplicity only one chain segment (bounded by curved arrows) rotates, its movement being compensated for by displacement of the plasticizer blob in the opposite direction (indicated by straight arrows). It is evident that the plasticizer blobs act like additional void volumes, allowing chain segments to rotate into the space which they occupy before the liquid 'flows' into the volume from which the chain segment moved. By analogy with the arguments advanced in the last section in connection with chain ends, the interpretation of the lowering of T_g which follows the incorporation of plasticizer into a polymer is obvious. In the system of partially interlocking gear wheels which was advanced in Section 3.4 to provide an analogy to a system of polymer chain segments, the effect of the plasticizer coming between polymer chain segments corresponds to the disengagement of some of the gear wheels. This means that the more of the gear wheels can become immobile before the entire system seizes up. The corresponding situation within the polymer is that less thermal energy in total is required to keep segmental rotation going and consequently it undergoes the transition to the glassy form at a lower T_g after the plasticizer has been added. DOP in poly(vinyl chloride) is a *permanent plasticizer*, i.e. it will remain within the bulk forever essentially. More than three quarters of all substances produced globally as permanent plasticizers goes into poly(vinyl chloride) (PVC). Figure 4.1(b) reveals the presence of DOP in PVC by its characteristic infrared band. Analyses of items made of 'PVC' reveal contents of about 25 per cent in carrier bags and about 40 per cent in plastic toys for instance.

Temporary plasticization is also a phenomenon of usefulness, particularly when water acts as a temporary plasticizer for synthetic fibres. For example, polyamides such as nylon 6,6 absorb water during washing, action which is promoted by the hydrogen bonding which results between the water molecules in the blobs and the $>C=O$ and $>N-H$ groups along the chain (see Fig. 3.2(b)). The result is that the amorphous phase in the drawn fibres is plasticized and in the case of nylon 6,6 this depresses T_g from around 330 K (typical of the dry polymer) to below ambient temperature. Thus flexibility is induced in the fibres to the extent that segmental rotations take out undesired creases in garments such as 'drip-dry' shirts when they are hung up to dry. Subsequently as the material dries out, T_g rises and on passing through ambient imparts stiffness to the fibres which gives the 'ironed' appearance.

When the polymer is composed of chains with substantial degrees of branching, there are two features of difference compared to the corresponding polymer composed of linear chains which are significant for T_g. The first is the presence in the backbone of $C-C$ links bearing three polymer chain substituents, as represented in Fig. 8.2(c), and the second is the additional chain end. It is the latter which in general appears to exert the greater influence, so that T_g is lowered by branching in the chains. But the effect is often overwhelmed by other differences, such as enhanced degrees of crystallinity of the linear chain polymer.

It is obvious that crosslinking produces the situation shown in Fig. 8.2(c) without the offsetting effect of additional chain ends. Hence T_g is increased by crosslinking because its only effect is to hinder segmental rotations in the vicinity of the junctions of crosslinks with the main chains. One instance of a simple crosslinked structure arises when a relatively small amount of 1,4-divinylbenzene is copolymerized with styrene to produce the structure shown in part in Fig. 8.4. The residues of single molecules of 1,4-divinylbenzene are the crosslinks between polystyrene chains effectively. When a parameter F is equal to the fraction of carbon atoms in the main chains which act as an anchoring point for a crosslink (examples starred in Fig. 8.4), the variation of T_g with F can be summarized by $T_g/K = 373 + 6.0 \times 10^2 F$. The upper limit of applicability of this equation is $F = 0.20$ when T_g has risen to 493 K, approaching the thermal decomposition range of this polymer. It is worth mentioning in connection

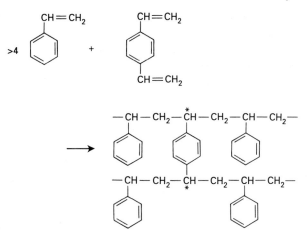

Fig. 8.4. Representation of the formation of a crosslink between polystyrene chains using divinyl benzene (asterisks mark the ends of the crosslink).

with styrene–divinylbenzene copolymers that these are commonly used as the packings for gel permeation chromatography columns (Section 2.2.2).

8.3 Drawn thermoplastics

To be useful for the manufacture of textiles, a fibre must have relatively high tensile strength and stiffness. These basic requirements are achieved in polymeric fibres usually when the polymer chain is very regular and there is strong interchain cohesion which promote high degrees of crystallinity. To give the fibre tensile strength, the crystallites are orientated by drawing (Section 3.2), which immediately places a restriction on the acceptable values of T_g. Cold drawing needs to be conducted at temperatures above T_g. But equally T_g needs to be well above ambient temperatures so that intentional creases, created by say ironing, are retained when the clothing incorporating the 'glassy' fibres is worn. A restriction on the value of T_m is imposed by ironing: the crystallites in the fibres must not melt at the ironing temperature, typically around 500 K, or otherwise the fibres would simply shrivel up as they lost the reinforcement of these solid regions. The creation of filaments by melt-spinning, as is common with nylon 6,6 and PET (Sections 5.3.1 and 5.3.2) requires that the polymer leaves the reactor at temperatures above T_m. Thus T_m must not exceed around 600 K or spinning will become a difficult and costly operation. Thus the synthetic polymers which have achieved main usage as textile fibres (Fig. 1.6) have the expected features of chains which are highly linear and amenable to close packing in a crystalline lattice. A generally stiff chain structure (as with poly(ethylene terephthalate) (PET) or hydrogen bonding between chains (as with nylon 6,6) promote crystallinity and hence fibre-forming ability.

It is very interesting to discover that apparently minor changes in the chain structures of both nylon 6,6 and PET can destroy their fibre-forming abilities. In the case of nylon 6,6, if the interchain hydrogen bonding (Fig. 3.2(b)) is disrupted by replacing some of the hexamethylenediamine residues ($-NH-(CH_2)_6-NH-$) by those of a dimethylhexamethylenediamine ($-NH-CH(CH_3)(CH_2)_4CH(CH_3)-NH-$) say, the result is an elastomeric material which does not crystallize on stretching (drawing), softens at relatively low temperatures, and readily dissolves into organic solvents, all in contrast with the properties of nylon 6,6. The additional methyl groups seem to prevent the chains from packing together closely enough to give a crystalline lattice with high cohesive energy. Similar effects result when the hydrogen atoms in the amide linkages ($-C(=O)-N(H)-$) of nylon 6,6 are progressively replaced by methylol groups to give linkages represented by ($-C(=O)-N(CH_2OH)-$). This prevents the formation of interchain hydrogen bonds at the chain separation represented in Fig. 3.2(b). The data in Table 8.5 indicates the progressive lowering of T_m and the change in the nature of the bulk polymer which accompanies this second substitution.

Table 8.5 Changes in T_m and the bulk nature of nylon 6,6 for various fractional replacements of H by CH_2OH in the amide linkages

Percentage of H replaced	T_m (K)	Bulk nature (~300 K)
0	540	Fibre-forming solid
25	478	Fibre-forming solid
60	295	Elastomer
75	227	Liquid

Similarly with PET, substitution of the chains can exert a dramatic change. For instance, when the glycol residue ($-O-CH_2-CH_2-O-$) has a methyl group substituted for one of the hydrogen atoms ($-O-CH_2-CH(CH_3)-O-$), the resulting polymer, poly(propylene terephthalate) (PPT), is a wholly amorphous polymer in bulk. One key factor in this dramatic loss of crystallinity compared to PET reflects the fact that propylene glycol ($HOCH_2-CH(CH_3)OH$), one of the monomers in the synthesis of PPT, has one OH group (on the left in the formula above) attached to a primary carbon atom whilst the other is bonded to a secondary carbon atom. This means that there are two possible configurations of each glycol residue in PPT, in CRU structures represented $-C_6H_4-C(=O)-O-CH_2-CH(CH_3)-O-C(=O)-$ and $-C_6H_4-C(=O)-O-CH(CH_3)$

$-CH_2-O-C(=O)-$. The random occurrence of these together with the relative bulk of the methyl group in PPT compared to a hydrogen atom in PET produces sufficient irregularity in PPT to inhibit crystallinity.

Fibres are produced by drawing of a suitable polymer in one dimension. Evidently drawing in two dimensions can result in a sheet: such material often appears in the windows of envelopes used to send bills or circulars. Drawing can also be used to make shaped containers of great strength for their weight, such as the large bottles for soft drinks which are usually made of PET. The basic shape is produced by injection moulding and this is drawn by blow moulding, which draws the bottle both lengthwise and round its short circumference (or hoop). These operations are represented in Fig. 8.5, which indicates typical draw ratios. The resulting wall thickness is only around 0.5 mm, but is nevertheless strong enough to form a bottle holding several litres. A similar bottle made of glass would be too weak to be of commercial acceptability.

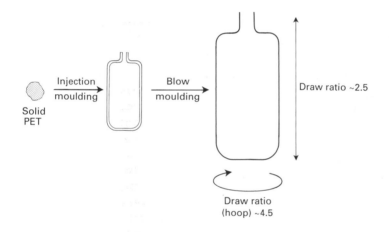

Fig. 8.5. Representation of typical operations involved in the manufacture of a PET bottle.

8.4 Elastomers

The common elastomers are rubbers, both natural and synthetic (Table 3.2), which have been lightly crosslinked, using residual $C=C$ in the monomer residues, to prevent chain slippage.

The most characteristic property of an elastomer is its ability to be stretched (or otherwise deformed) but to regain its original shape rapidly (or 'snap back') on release of the stress. In this it acts as an 'entropy spring' rather than as an 'energy spring' of which a steel spring is a typical example.

The essential microscopic features of a rubber are the ease of segmental rotation down to rather low temperatures (i.e. a low T_g) and the virtual absence of (micro)crystallinity. These imply low energies of cohesion of chains and hence rather small enthalpy changes (ΔH) when the chains are uncoiled, as occurs when the rubber is stretched. Consider at first a single unbranched chain: at any instant there will be a particular separation (l) between the two ends and l will vary from one instant to the next as segmental rotations occur within the chain. There will be a maximum value of l (denoted as l_{max}) which is the limit which can be achieved without energy considerations becoming of main significance. The situation in which $l = l_{max}$ is evidently produced in poly(cis-1,4-isoprene) (natural rubber) for example when the entire chain has the zigzag conformation shown in Figure 3.9. The minimum value of l is zero, achieved when the chain ends touch, a situation which is evidently of very low probability compared to intermediate values of l. One of the intermediate values will be that for which the number of conformations (W_l) is a maximum: this value of l is also the most probable end-to-end separation for a large number of chains and accordingly is denoted by l_{mp}. There is a fundamental statistical thermodynamic relationship between the entropy (S) of a system and the number of distinguishable arrangements which can be produced within it. In the present context this yields $S_l \propto \ln W_l$. The total entropy of an assembly of many chains is expressed by $S_T = \int S_l dl$, where the entropy contributions (S_l) from all chains of end-to-end length l are integrated over all possible values of l, i.e. from 0 to l_{max}. The full line profile shown in Fig. 8.6 corresponds to the equilibrium state of an

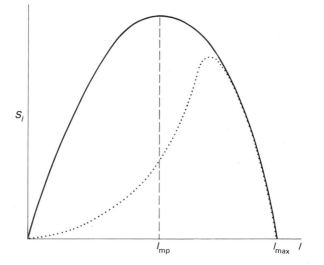

Fig. 8.6. General forms of plots of S_l versus l for a piece of an elastomer (rubber) when it is unstretched (continuous profile) and after stretching (dotted profile).

unstretched piece of rubber in general terms.

When this rubber is stretched, the chains must adopt more elongated conformations to relieve the applied tensile stress. This decreases the probability and hence the value of W_l of lower values of l, i.e. S_l is lower for the stretched rubber on the left-hand side of Fig. 8.6 but is unaltered on the right as is indicated by the dotted profile. The envelope enclosed by the dotted profile measures S_T for the stretched rubber and it is obvious that the total entropy decreases as the piece of rubber is stretched. It is then the gain in total entropy which accompanies the retraction of the rubber on its release which provides the driving force. The rubber snaps back quickly in reflection of the rapidity of the segmental rotations which take place at the microscopic level to produce the retraction at the macroscopic level.

The fundamental requirements in an elastomer are highly mobile chain segments with anchoring sites at which chains are prevented from slipping past one another. Vulcanization achieves the latter by chemical bonds (Fig. 3.4), to produce an elastomer which cannot be remoulded into a different physical form because of the permanence of these interchain links. Conventional rubbers have to be cut to shape usually, involving considerable wastage of material. This disadvantage is overcome when *thermoplastic elastomers* are used. These have physical rather than chemical links between chains and thus, as the name implies, can be melted and moulded to shape, many times if desired. What is required usually is a block copolymer in which the blocks have very different natures. One block is a typical elastomeric chain, very mobile via segmental rotation but with no tendency towards crystallinity. The other block is the chain of a thermoplastic, with a strong tendency towards crystallinity. The most common examples have chains which are ABA triblock copolymers, with B representing a poly(*cis*-1,4-butadiene) elastomeric block and A representing relatively short chains of highly stereoregular polystyrene, often referred to as SBS polymers. The key point is that polybutadiene and polystyrene as separate polymers are incompatible so that they are disinclined to mix together. In the SBS triblock copolymer the S blocks tend to come together to form *domains* which are small regions with high degrees of crystallinity embedded in the amorphous mass created by the agglomeration of the B blocks. The domains have relatively high thermal stability and serve as the anchoring points for the B chains, in which there is rapid segmental rotation at relatively low temperatures. Figure 4.12 revealed the existence of such domains in the SAXS profile for a similar triblock copolymer. The glass transition temperatures (T_g) of polystyrene and polybutadiene are ~ 365 K and ~ 215 K respectively. Therefore at temperatures below 215 K, the SBS polymer is a hard glass, since the polybutadiene chain segments are immobile. Between 215 K and 365 K, a range which includes the temperatures of normal usage, SBS polymer behaves as an elastomer with the polybutadiene chain segments highly mobile but prevented from slippage by the polystyrene domains which are rigid (being below T_g) and serve in a similar way to the crosslinks in vulcanized rubber. Above 365 K the styrene domains will become flexible. But when the temperature is raised above the T_m of stereoregular polystyrene (~513 K for the isotactic and ~542 K for the syndiotactic forms), the domains cease to exist and the molten polymer can be injection moulded into the shape required, for example soles for shoes. SBS triblock copolymer is synthesized by anionic living polymerization (Fig. 7.3) and it and a similar SIS (styrene–isoprene–styrene) triblock copolymer have been marketed under trade names of Kratons.

The second commonest type of thermoplastic elastomers have polyurethane structures (Fig. 1.8). Their nature stems from their chains being composed of alternating blocks of a so-called 'soft' type (having easy segmental rotations at ambient temperature) and a 'hard' type which combine with others through strong mutual attractions to form domains. The first step of synthesis of these thermoplastic elastomers is the production of a prepolymer in which the chains are terminated by isocyanate groups ($-N=C=O$). This usually involves the use of a preformed chain terminated by $-OH$ groups, typically a polyester glycol or a polyether glycol with $\bar{M}_n$ of 2000–3000, which is reacted with a diisocyanate (typically methylene-4,4'-diphenyl isocyanate (MDI)) as represented in Fig. 8.7. The resulting prepolymer is reacted with a diamine, typically ethylenediamine, to link the chains and thus generate the chains of the thermoplastic elastomer.

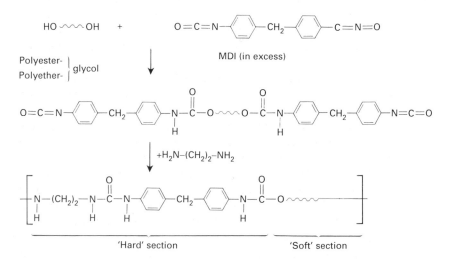

Fig. 8.7. Representation of the scheme for synthesis of a polyurethane type of thermoplastic elastomer.

It is the mobility of the 'soft' sections (indicated by a wavy line) of the chains which produces the elastomeric nature of the polymer. The 'hard' sections are created by the urethane segments and these are involved in interchain hydrogen bonding ($>C=O-$ $-H-N<$) which is a principal driving force for these sections to agglomerate in the domains which stop chain slippage. These segmented polyurethanes have the general chain structure $(A_x B_y)_n$, in which the hard (say A_x) and soft (B_y) sections are relatively short blocks which alternate n times. Commercial thermoplastic elastomers of this type are often used in fibre form, a common example often going under the trade name of Spandex.

'Star' polymers have molecules which are composed of three or more polymeric chains extending from a central structural unit. Some star polymers may be more explicitly described as *radial block copolymers* and amongst these are some that show elastomeric properties in bulk. It is these which are of specific interest here, leaving further aspects of star polymers to be discussed in Chapter 10.

A common synthetic route to star polymers having a few arms only involves the use of silicon tetrachloride ($SiCl_4$) or a related compound to couple the living chain ends in an anionic polymerization system (Section 7.3). This procedure results in the elimination of lithium chloride (LiCl) when an alkyl lithium initiator has been used, leaving the grown chains bonded to the silicon atom in place of the original chlorine atoms. A case of interest used styrene (S) and butadiene (B) as the monomers and secondary-butyl lithium as the initiator, with polymerization conducted in solution in cyclohexane at ~ 273 K. The monomers were added in turn, to give either $S_m B_n^-$ or $B_m S_n^-$ diblocks with living ends depending on the order of addition of the monomers. In this case the coupling agent was 1,2-bis(trichlorosilylethane), which, has the formula $H_2C(SiCl_3)-C(SiCl_3)H_2$ and can give six-armed stars in principle, in which each arm is a diblock copolymer. Two distinguishable types of polymer arise, reflecting the order in which the monomers are added. When styrene is added first, the final star polymer has polystyrene blocks as the outer parts of the arms and is labelled as SBH. The reverse order of addition gives outer blocks of polybutadiene and this polymer is labelled as BSH. Films of these polymers were cast from solutions in two different solvents. One, cyclohexane (Cy), dissolves polybutadiene much more readily than polystyrene. The other solvent (T/M) was a 9:1 v/v mixture of tetrahydrofuran and methylethylketone; polybutadiene and polystyrene are both easily soluble in this. The resultant films were examined using a transmission electron microscope and Fig. 8.8 shows the spectacularly different microstructures, when the regions composed of polybutadiene blocks have been stained using osmium tetroxide and appear black. Figure 8.8(a) shows an ordered arrangement of domains of B blocks which appear as dots of around 15 nm diameter in the white matrix of S blocks. This is to be regarded as the 'normal' morphology of BSH since the film is cast from T/M solution in which the B and S blocks are both very soluble. The separation of the blocks is driven by their incompatibility, just as was the case for the SBS triblock copolymer with linear chains discussed before. It is because the B blocks are the outer parts of the star arms in BSH that they can aggregate into domains. In contrast in the film of SBH cast from T/M which is seen in Fig. 8.8(c), there is incomplete separation. In SBH the butadiene blocks are inside S blocks in the arms of the star polymer so that their ability to form domains is limited and the result is the disordered array. The greyish regions are created by very small particles composed of B blocks in the 'core' of a single polymer molecule dispersed fairly uniformly through a phase composed of S blocks: these zones may be regarded as 'well-mixed' in terms of the two types of blocks and this illustrates the ability of block copolymerization to bring incompatible polymer chains into intimate contact, more than just at the junctions between blocks in the individual polymer molecules.

When the discriminating solvent Cy is used to cast the films, they show the completely different morphologies evident in photographs (b) and (d) of Fig. 8.8. Lamellar microstructures are apparent with both BSH and SBH polymers, corresponding to layering of the two types of block. These morphologies come about simply because they are the most stable ones as the casting solutions separate into a solid phase composed of relatively insoluble S blocks with the B blocks tending to remain in solution.

The mechanical behaviours of these BSH and SBH polymers are of interest. Key points are the glass transition temperatures of ~373 K for stereoregular polystyrenes (both isotactic and syndiotactic forms) and <210 K for the geometrical isomers of polybutadiene (Table 3.3 for example). Thus B blocks have easy segmental rotations whilst S blocks form rigid parts of the structures at normal temperatures. The star polymeric films cast from T/M solutions have been found to break relatively easily under tensile stress, whilst of the films cast from Cy solutions SBH is considerable stronger than BSH. The key feature at the microscopic level in these Cy-cast films is the continuity of the lamellar domains created by B blocks which have the nature of elastomers and thus relieve strain. In T/M-cast films it is the glassy phase created by agglomerated S blocks which is continuous whilst the discontinuity of the phase created by the B blocks does not allow the resultantly rather brittle material to avoid rupture under modest strains (Table 8.2). The relative weakness of BSH compared to SBH reflects the fact that in the former it is the lamellae composed of B blocks which contain the ends of the arms of the stars. A chain end in an amorphous elastomeric phase can be expected to expand the structure in its vicinity (Section 8.2) so that overall the chains are less tightly entangled with one another. The polymeric film in which the lamellae are composed of the more expanded polybutadiene phase (i.e. BSH) would be expected to be weaker than the film not only composed of denser

(a)   (b)

(c)   (d)

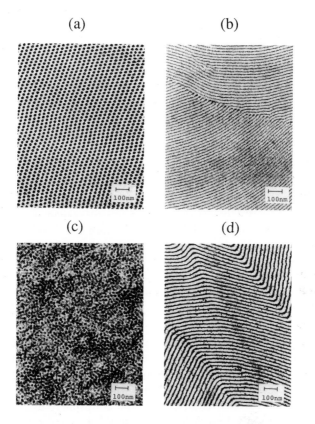

Fig. 8.8. Transmission electron microscope (TEM) photographs of BSH- and SBH-star polymer films cast using two different solvents. (a) BSH film cast from T/M solvent, (b) BSH film cast from Cy, (c) SBH film cast from T/M, and (d) SBH cast from Cy. Reproduced with permission of the Society of Polymer Science, Japan from Takigawa, T., Ohta, Y., Ichikawa, S., Kojima, T., Tanaka, A., and Masuda, T. (1988). Morphology and viscoelastic properties of star-shaped styrene–Butadiene radial block copolymers. *Polymer Journal*, **20**, 293–305.

polybutadiene but in which the chain ends are locked into the glassy polystyrene lamellae.

None of the materials concerned in Fig. 8.8 is a thermoplastic elastomer in fact: that would demand the reverse situation to that seen in Fig. 8.8(a), in which domains of polystyrene were embedded in a continuous polybutadiene phase. Nevertheless, the cases of SBH and BSH considered provide a nice illustration of domain formation and show how constitution and pretreatment of such polymers can govern the physical nature of the resultant solid.

Further instances of behaviour corresponding to that expected of thermoplastic elastomers will be encountered in Section 9.5 in connection with ionomers.

8.5 Polymer degradation and recycling

There are some circumstances in which polymeric materials must be highly durable, such as the poly(vinyl chloride) used in both the sheathing of electrical wires and in the manufacture of what are often known as UPVC window and door frames. But in many other uses, it is desirable that the polymers should degrade over a period of time to avoid creating an environmental problem. In general, incineration and recycling are the only viable schemes.

Recycling without chemical alteration of the polymer is only possible in a straightforward manner with thermoplastics. Elastomers with chemical crosslinking and thermoset resins with gigantic network structures will plainly be difficult to deal with. Additionally much will depend on the overall composition of the bulk material, since the presence of additives or more than a single polymer will pose problems of separation. In general what is required is a homogeneous thermoplastic material. In practice there must be a considerable potential for increased recycling of plastics, when less than 6 per cent of the polymeric materials which are amenable in this respect (much of it in the form of packaging) are being recycled in the USA at present for instance.

Almost all polymers incorporating only carbon, hydrogen, oxygen, and nitrogen can be incinerated successfully without special procedures. Problems can arise when other elements are present. A major problem is encountered in the incineration of poly(vinyl chloride), since this results in the release of hydrogen chloride into the atmosphere usually. It has been estimated that some 20 million kilogrammes of this gas is discharged into the air above Western Europe as a result of the incineration of poly(vinyl chloride). But this is put into perspective when it is realized that the total emission of hydrogen chloride which results from the combustion of coal is about an order of magnitude larger. There may still be localized problems in the vicinity of the incinerator, because hydrogen chloride is so soluble in water that much of the subsequent deposition of hydrochloric acid may well take place close to the emission source. The plastics fraction extracted from municipal waste will be composed of several plastics: by weight, 58 per cent polyalkenes (mainly polyethylene), 15 per cent poly(vinyl chloride), and 20 per cent of polystyrene is a representative composition of the major components. It is uneconomic to separate the poly(vinyl chloride) from the other polymers in this waste, so it all tends to be incinerated.

Thermal degradation or pyrolysis offers some practical possibilities for the reclamation of substances of value from polymeric waste. In some cases, the resultant production of monomer is sufficient to be commercially viable as a depolymerization/repolymerization mode of recycling. When polystyrene is heated at 573 K under nitrogen or even air, the yields of styrene at ~30 per cent w/w are the largest of a single substance by at least

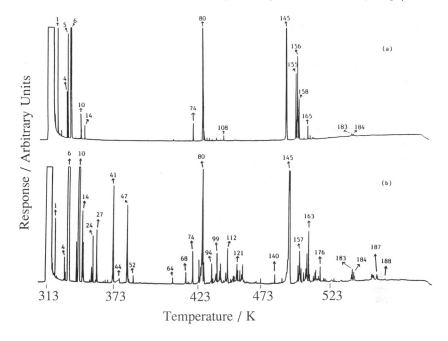

Fig. 8.9. Gas chromatograms showing peaks corresponding to volatile compounds produced during the thermal decomposition of polystyrene in (a) nitrogen and (b) air at 573 K. Identification of major peaks is given in the text. Reproduced with permission of Elsevier Science Publishers BV from Shappi, M. M. and Hesso, A. (1990). Thermal decomposition of polystyrene: volatile compounds from large-scale pyrolysis. *Journal of Analytical and Applied Pyrolysis*, **18**, 143–61.

a factor of two. Many other compounds are evolved in minor amounts, as is indicated in Fig. 8.9. Styrene (labelled 6) appears early in the chromatograms as the column temperature of the gas chromatograph was raised from 40°C (313 K) at a rate of 5 K min^{-1}. Benzaldehyde (labelled 10) is the next most abundant product when air is present during the thermal decomposition. However the recovery of the styrene and purification to a polymerization grade monomer would be expensive to bring in on a large scale, so that this may not be the basis of an industrial operation for recycling polystyrene.

The greatest potential for pyrolysis as a means of recycling thermoplastics probably lies with poly(methylmethacrylate) (PMMA) and poly(α-methyl styrene) when depolymerization to yield the monomer is the overwhelming process. For instance it has been demonstrated that pyrolysis of PMMA in a fluidized-bed system at $\sim$720 K can result in the recovery of more than 97 per cent of the mass as the monomer (MMA). Even PMMA obviously containing additives, such as the coloured rear light casings of vehicles, can be effectively recovered in this way.

For thermal decomposition of a thermoplastic to yield the monomer almost exclusively, it seems to be necessary that the monomer concerned is a 1,1-disubstituted vinyl compound. When the polymer is heated, a key feature of the mechanism of depropagation (or 'unzipping' more descriptively) is that when a monomer residue is detached from the end of the chain by breaking a $C-C$ bond, the residual chain then has a tertiary radical site at its end. The relatively high thermodynamic stability of such a tertiary site seems to have the effect of suppressing any potential for scissions of $C-C$ bonds to occur anywhere in the backbone apart from at the end of the chain. In fact in the range above 540 K in which the thermal decomposition of PMMA proceeds at a significant rate, the detachment of single monomer residues from the end of the unzipping chain seems to be positively encouraged.

From the point of view of protecting the environment, it would be desirable for most polymeric materials to be biodegradable; this would demand that they must decompose almost completely under the attack of microorganisms or fungi over periods of months at the most. It is evident that most commercial plastics are to be regarded as nonbiodegradable, when we see the amounts of discarded plastics which litter our cities, countryside, and beaches today. Perhaps the worst polymers in this respect are polypropylene, PET, polystyrene and poly(vinyl chloride) since they are virtually immune to attack by the enzymes which are responsible for biological oxidation processes in general. Photodegradation of polymers under the action of sunlight seems to be a more realistic option in a few cases. One interesting instance concerns the polyethylene collars which are used to make packs (for example 6-packs) of cans of drinks. It has been found that if this polyethylene is synthesized by a free radical mechanism in the presence of carbon monoxide-gas, the result is a chain incorporating a few carbon monoxide residues. These copolymers of ethylene and carbon monoxide, with around 1 per cent of residues of the latter, undergo photodegradation in sunlight over a week or so. The normal commercial forms of polyethylene are not subject to any significant photodegradation in contrast. Otherwise, there have only been some rather unsuccessful attempts to achieve photodegradation of various plastics via the agencies of incorporated coloured cations, such as ferric (Fe^{3+}) and ceric (Ce^{4+}), or photosensitizing molecules, such as benzophenone.

On the other hand, the commercial importance of other uses of polymers depends on their ability to resist all forms of degradation over many years. The mechanisms through which polymers are altered over long periods of exposure to natural elements are therefore of considerable concern. Vinyl polymers are of particular interest, in view of the widespread usage of poly(vinyl chloride) in frames of windows and doors, often under the tradename of UPVC. Crosslinking tends to increase as time passes. The key initiating process appears to be the attack on $\geqslant C-H$ bonds by hydroxyl radicals (OH), the ubiquitous reactive species in sunlit air. Hydrogen abstraction forms water and leaves a radical site $\Rightarrow$C·; pairs of such sites are likely to combine to form $\Rightarrow$C—C$\Leftarrow$ bonds, i.e. crosslinks (Section 3.3). But the overall mechanism is highly complex and it is not possible to be definitive on the elementary processes which are of significance.

It is well known that polymers with C=C linkages in their backbones are much more susceptible to oxidation-induced degradation than are those with only C—C linkages. The relevant chemical actions are numerous and in general their full mechanisms have not always been established. But the points which can be made with certainty are that free radical sites are central features and that intermediate species of the nature of both peroxides (—O—O—) and hydroperoxides (—O—OH) are involved. Resistance in this respect is provided by substances referred to as *antioxidants* which are commonly incorporated into many polymeric materials. The effective action of the antioxidants is thought to depend upon either or both of the destruction of hydroperoxide groups and the annihilation of free radical centres which are central to the oxidative chain reactions. The most commonly used antioxidants are substituted phenols, aromatic amines, and various thio-compounds.

Natural rubbers and their synthetic analogues have many C=C bonds and degenerate (or 'perish') with exposure to light and air. Incorporated antioxidants, often phenylnaphthylamines, are able to extend greatly the useful lifetimes of both natural rubber and synthetic SBR varieties based upon styrene(S) and butadiene(B) residues. Perished rubber becomes brittle, having lost most of its initial strength and flexibility. Ozone (O_3) is a particular enemy of rubbers, to the extent that there is a separate class of additives known as *antiozonants*, exemplified by various derivative compounds of 1,4-diphenylenediamine. Ozone attacks the C=C linkages in the backbone via the usual ozonolysis process familiar with small alkenes, which ultimately cleaves the bond and gives rise to new chain ends terminated by carbonyl groups. Now it seems that the chains in which there is a methyl (—CH_3) side-group adjacent to the double bond are by far the most susceptible to oxidative degradation processes in general. Thus natural rubber, which is based on isoprene residues (usually —CH_2—C(CH_3)=CH—CH_2—), perishes rather quickly as found in practice. Further, it seems that such methyl side-groups also make the chains prone to scission and crosslinking; increasing degrees of the latter accompany the disappearance of C=C bonds and account for the decreasing flexibility of the rubber as it perishes. SBR (Section 6.2.4) is more resistant than natural rubber, simply because the butadiene residue (—CH_2—CH=CH—CH_2—) has no methyl side-group. Butyl rubber is the copolymer of isobutene with a very small fraction of isoprene, the latter giving rise to the residues containing C=C bonds, which subsequently are mostly used up in forming the crosslinked elastomer. Thus the predominant monomer residue in butyl rubber is $-C(CH_3)_2-CH_2-$ and the virtual absence of $C=C$ bonds gives it high resistance to perishing.

Exercises

8.1. WAXS (Section 4.3.1) has been used (as in connection with Fig. 4.7) to estimate the fractional degree of crystallinity (X per cent) of three samples of poly(etheretherketone) (PEEK). The densities (ρ) were also measured. The corresponding values are listed below

| X (%) | 18.4 | 27.4 | 35.0 |
| ρ (kg dm^{-3}) | 1.283 | 1.292 | 1.300 |

What are the values of (a) the maximum density and (b) the minimum density of PEEK on the assumption that the interphase between crystallites and the amorphous polymer is of negligible significance?
Source: Hay, J. N., Kemmish, D. J., Langford, J. I. and Rae, A. I. M. (1984). *Polymer Communication*, **25**, 175–8.

8.2. ^{13}C NMR spectra of three different linear low-density polyethylenes (LLDPEs) showed prominent peaks centred on the values of chemical shift (δ (ppm)) listed below

LLDPE 1 39.7 34.1 30.5 27.1 26.7 11.2
LLDPE 2 38.2 34.6 34.2 30.5 29.6 27.3 23.4 14.1
LLDPE 3 38.2 34.6 34.5 32.2 30.5 30.0 27.3 27.2 22.9 14.1

What are the likely structures of the chains?
Source: De Pooter, M., Smith, P. B., Dohrer, K. K., Bennett, K. F., Meadows, M. D., Smith, C. G. *et al.*, (1991). *Journal of Applied Polymer Science*, **42**, 399–408.

8.3. The plot of the specific volume (i.e volume per unit mass) of a typical elastomer vs temperature shows a distinct break at a point where a line of lesser gradient extending to lower temperatures intersects a line of steeper gradient extending to higher temperatures. The temperature at this point corresponds to the glass transition temperature (T_g). How may this be explained?

8.4. Ring chains (i.e. with no chain ends) and also in each case linear chains with the same *weight* average degree of polymerization ($\overline{DP}_w$) of polystyrene have been synthesized. The glass transition temperatures (T_g) were found to be as follows for corresponding samples

$\overline{DP}_w$	3000	200	70
T_g(K) (ring)	378	378	378
T_g(K) (linear)	378	372	359

(a) How may these data be interpreted?
(b) Devise an equation relating the value of T_g to $\bar{M}_w$ for the polystyrene samples composed of linear chains concerned here.
Source: Antonietti, M. and Fölsch, K.J. (1988). *Makromolekulare Chemie, Rapid Communications*, **9**, 423–30.

8.5. The glass transition temperature (T_g) of nylon 6,6 varies with its water content expressed in terms of number of moles of water per mole of constitutional repeating units (CRUs) (X_{water}), according to the linear equation

$$T_g(K) = 354 - 170X_{water}.$$

Given that nylon 6,6 can absorb up to 7 per cent w/w of water, what is the general significance of this equation?

Source: Birkinshaw, C., Buggy, M. and Daly, S. (1987). *Polymer Communications*, **28**, 286-8.

8.6. Figure 8.10 shows differential scanning calorimetry (DSC) scans of poly(ethylene terephthalate) (PET) fibres at various extents of elongation (up to 120 per cent of original length) by drawing (NDR indicates the 'natural draw ratio', corresponding to 35 per cent elongation). The cold crystallization temperature for the as-spun PET filament is 110°C (383 K) and the glass transition temperature is 67°C (340 K). What may be deduced from comparison of the various profiles? (Fig. 3.11 may be referred to with advantage).

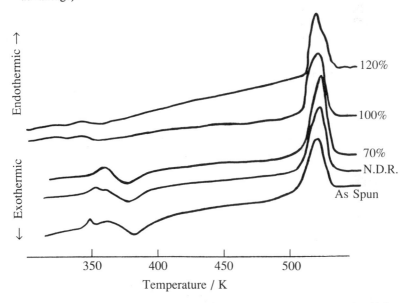

Fig. 8.10. Differential scanning calorimetry (DSC) profiles of a spun and cold-drawn fibre of poly(ethylene terephthalate) (percentage degree of elongation indicated on the right). Reprinted by permission of John Wiley & Sons, Inc. from Napolitano, M. J. and Moet, A. (1987). Mechanism of cold drawing in melt-spun poly (ethylene terephthalate) fibers. *Journal of Applied Polymer Science*, **34**, 1285-1300. Copyright (1987) John Wiley & Sons, Inc.)

8.7. A rubber band suspended by a clamp and stretched by an attached weight is found to contract in length when warmed (using a hot-air blower say). How may this be explained?

8.8. A multiblock copolymer with chains composed of alternating polystyrene and poly(propylene oxide) blocks has been synthesized, so that every block of each type has approximately the same length. The polystyrene blocks (prepolymers) were prepared by anionic polymerization to have $\bar{M}_n$ equal to 1.14×10^4. The poly(propylene oxide) blocks (prepolymers) have $\bar{M}_n$ of 3.12×10^4. A copolymer which incorporated these blocks had $\bar{M}_n = 15.0 \times 10^4$: a strip of it could be stretched at ambient temperature to about eight times its original length with ease and snapped back on release. This copolymer was a rigid solid at temperatures below 226 K, was a rubbery solid up to 381 K and softened noticeably above that temperature to melt at approximately 520 K.

(a) What are the average numbers of monomer residues in each of the prepolymers?

(b) Deduce the most probable chain structure of the copolymer.

(c) What are the relative weight ratios of the two types of blocks in the structure devised in (b)?

(d) How may the specified mechanical properties of the copolymer be explained? (Refer to Tables 3.1 and 3.3.)

Source: Xie, H. and Chen, X. (1988). *Makromolekulare Chemie, Macromolecular Symposia*, **20/21**, 19–28.

8.9. Explain how Spandex (see in vicinity of Fig. 8.7 in text) can be drawn into a fibre at elevated temperature and how the resultant fibre is elastic at ambient temperature.

9
Some speciality polymers

This chapter will look at some polymers which possess useful properties even if they are not produced in amounts approaching the scale of the polymers concerned in Chapter 8. Inevitably there will be a somewhat indistinct division between this and the next chapter, which attempts to look into the future.

9.1 Graft copolymers and comb copolymers

In connection with copolymers, a graft can be defined generally as a polymeric chain which is attached via a chemical bond directly to the backbone of a different polymer at a point other than an end. Some remarks on grafting and the syntheses of graft copolymers have been made earlier (Section 6.6). The common forms of graft copolymers have the junctions of the attached chains spaced along the backbone at random intervals.

Grafting can often be used to improve or modify particular properties of synthetic polymers. Section 6.6 includes the description of the improvement of the impact resistance of poly(vinyl chloride) when chains of polybutadiene are grafted on, thus combining thermoplastic and rubbery natures respectively and leading to the formation of domains at the microscopic level. A similar view may be taken of the 'shatterproof' nature of ABS copolymers, in which thermoplastic chains of acrylonitrile and styrene have been grafted onto rubbery polybutadiene. Another case concerns isotactic polypropylene, which as a pure polymer is inert towards many chemical reagents to the extent that it is difficult to dye fibres, print onto sheets, and to achieve strong adhesion in general. These difficulties are overcome if 4-vinylpyridine or acrylic acid chains are grafted on. Such grafting has been initiated by subjecting polypropylene to γ-radiation from a ^{60}Co source in air, which is believed to produce hydroperoxide ($-$OOH) groups. These are subsequently thermally decomposed to yield radical sites on the polymer surface in the presence of the monomer which will be incorporated into the grafted chains.

A spectacular exemplification of a change in a polymer surface effected by grafting is shown by the photographs in Fig. 9.1. These show the result of placing a droplet of water on a styrene–butadiene copolymer surface.

Fig. 9.1. (a) A droplet of water placed on a film of ordinary styrene–butadiene ruɒber. (b) The film of water indicating the wettability of the underlying hydrophilic polymer film produced by grafting of a copolymer including blocks of poly(ethylene oxide). Reprinted by permission from Noda, I. (1991). Latex elastomer with a permanently hydrophilic surface. *Nature*, **350**, 143–4. Copyright (1991) Macmillan Magazines Ltd.

Ordinarily the hydrophobic nature of the polymer gives rise to the bead of water shown in Fig. 9.1(a). But when chains of poly(ethylene oxide) have been grafted, their incompatibility with the main chains results in their accumulation at the surface, making it relatively hydrophilic. This is manifested in Fig. 9.1(b) which shows the spread of the water droplet on the surface of the grafted copolymer.

A *comb (co)polymer* may be regarded as a very regular type of graft polymer in which the junctions are separated by uniform short lengths of backbone, when special properties may result. Perhaps the simplest way in

which the synthesis of a comb-like chain is achieved is by the polymerizations of a vinyl monomer represented as $CH_2=CH(R)$ or an alkyl
methacrylate monomer $CH_2=C(CH_3)(C(=O)OR)$, in which R is an alkyl
chain. A first point of interest is how the glass transition temperatures
(T_g) of the resultant polymers change as R becomes longer. Normally T_g
would be expected to rise with the increase in the size of the substituent
group because of the associated increase in the energy barrier to internal
rotation about the backbone $C-C$ bond (Section 3.4). But in these cases
there is a decrease of T_g as R lengthens, as for example with the poly-
methacrylates having $R = (CH_2)_n CH_3$ with n up to 11. It is believed that
a phenomenon known as internal plasticization operates. The n-alkyl chains
themselves are very flexible and can thus move easily to create space for
segmental rotations in the backbone. The longer is the chain represented
by R, the larger is the volume occupied by the flexible 'teeth' of the combs
(i.e. the R chains) and the more separated are the backbones. Thus the R
groups act at the microscopic level in a manner not unlike the molecules
of the liquids normally used to effect plasticization, as shown in Fig. 8.3.

Even more extraordinary behaviour may be observed in connection with
crystallinity and hence melting phenomena. Figure 9.2 shows differential

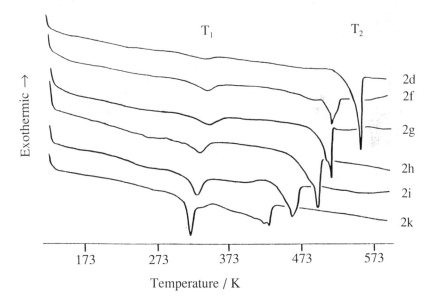

Fig. 9.2. Differential scanning calorimetry (DSC) profiles of substituted poly(*p*-pheny-
lene) derivatives with *para* substituent groups $-(CH_2)_n CH_3$, with $n = 5$ (2d), 6 (2f), 7
(2g), 8 (2h), 11 (2i), and 15(2k). Transitions are observed at temperatures of T_1 (lower
temperature) and T_2. Reproduced with permission of Hüthig & Wepf Verlag, Basel, from
Rehahn, M., Schlüter, A-D., and Wegner, G. (1990). Soluble poly(*para*-phenylene)s, 3.
Variation of the length and the density of the solubilizing side chains. *Makromolekulare
Chemie*, **191**, 1991–2003.

scanning calorimetric (DSC) profiles (Section 3.6) of a series of poly(p-phenylene)s, in which the aromatic rings are substituted at positions 2 and 5 with identical R groups. The chain can be regarded as a double comb with a rather stiff but highly regular backbone in each of these polymers. Two distinct melting points, designated as T_1 and T_2, are evident, with T_1 becoming more pronounced and T_2 becoming significantly lower as the n-alkyl chain represented by R lengthens. In crystalline regions of one of these polymers at ambient temperature or below, not only are the main chains composed of aromatic rings aligned in a lattice but so also are the side chains (R), the more so the longer they are. T_1 is associated with the melting of the side chains only, so that although the main chains remain fixed between T_1 and T_2, these R chains become mobile within the intervening space. The endothermicity (as indicated by the depth of the trough in the trace at T_1) increases as the R chain lengthens, which is expected as each of its component $-CH_2-$ groups makes a contribution to the cohesive energy concerned. At T_2 the phenomenon is the melting of the main chain. The decrease of the value of T_2 as R lengthens may be ascribed simply to the increasing separation of the main chains, which would be expected to reduce the interchain cohesive energy. But with some other polymers, such as poly(alkyl methacrylate)s, only a single peak appears in each DSC profile: this is observed for example with the polymer having the monomer residue represented as $-CH_2-C(CH_3)(C(=O)O-CH_2-(CH_2)_{16}CH_3)-$. The phenomenon of side-chain crystallinity (giving rise to T_1) in the absence of main-chain crystallinity (i.e. no T_2) is believed to provide the interpretation of this behaviour.

Comb-like polymers can also exhibit the phenomenon known as liquid crystallinity, the topic developed in the next section.

9.2 Liquid crystal polymers

The individual molecules in a liquid crystal have a restricted mobility which preserves a degree of order compared to a true liquid. The consequence is that liquid crystals show *anisotropic* behaviour, which means that their properties vary with the direction in which they are measured, unlike the *isotropic* natures of true liquids. Liquid crystals are often termed *mesophases* because they show behaviour which is apparently intermediate between that of a crystalline solid and a normal liquid. The small (i.e. non-polymeric) molecules which exhibit liquid crystallinity tend to be rather rigid with rod- or disc-like shapes. The phenomenon originates in forms of long-range order extending over many millions of molecules, so that within a small region there is a *director* which is manifested in the preferred alignment of the molecules on a statistical basis. Thus although the individual molecules are in constant motion, as are those in a liquid, there is a rapid fall-off in probability that the axis (or plane) of any particular

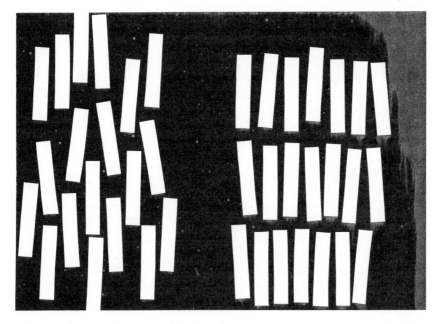

Fig. 9.3. General representation of nematic (left) and smectic (right) forms of liquid crystallinity.

molecule will deviate far in terms of the angle from the most probable direction at any instant. But on the larger scale, the directors in well-separated regions of a bulk sample lie at random angles with respect to one another, in the absence of any aligning influences such as electrical fields, etc.

There are two types of microscopic ordering which are of main interest here; these are illustrated in Fig. 9.3 using rectangular blocks to represent stiff elongated sections of molecules. In the *nematic* form there is only longitudinal orientation, but in the *smectic* form there is also a pronounced layering tendency. There are two primary classes of liquid crystal which indicate the method of creation: the *lyotropic* type forms an ordered solution above some critical concentration in a suitable solvent and below a critical temperature whereas a *thermotropic* liquid crystal is produced in the melt of the pure substance.

The long chains of polymer molecules might seem to be compatible with liquid crystallinity. But most are too flexible and, as a result of entanglements and coiling of chains, they cannot achieve the degree of statistical orientation over relatively long ranges which is required. It is only when the polymer has a chain which is rather stiff and thus has very limited mobility that many chains can be aligned relatively easily in a bulk sample. The director becomes particularly effective when the chains are also attracted together via hydrogen bonding or coupling of induced dipoles, the

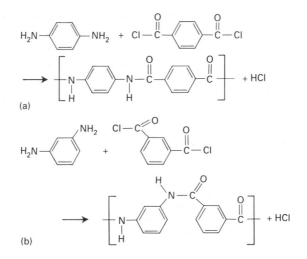

Fig. 9.4. Simplified representation of the conversions of the monomers into the CRUs in the common syntheses of the aramids Kevlar (*top*) and Nomex (*bottom*).

latter being enhanced when aromatic rings and multiple bonds are present in the chains.

Perhaps the best known polymeric material which relies upon the phenomenon of liquid crystallinity in its production is poly(*p*-phenylene terephthalamide) or Kevlar as it is known commercially. Kevlar is an aromatic polyamide (or aramid in generic terminology) with the structure of its constitutional repeating unit (CRU) shown in Fig. 9.4. Aramids are synthesized generally by step-growth polymerizations, usually with the appropriate aromatic diamine and aromatic acid dichloride as the monomers which are dissolved in the typical amide solvent. Figure 9.4 indicates the synthetic routes to Kevlar (a) and Nomex (poly(*m*-phenylene isophthalamide)(b).

The secret of the extraordinary properties of Kevlar comes from the lyotropic liquid crystalline mesophase which is produced when the polymer is dissolved (15–20 percent w/w) in concentrated (100 per cent nominal) sulphuric acid. The usage of this unpleasant solvent creates many technical difficulties, but poly(*p*-phenylene terephthalamide) is just not sufficiently soluble in common organic solvents. In the solution, the chains are strongly aligned, aided by intermolecular $C=O---H-N$ hydrogen bonds. Moreover the directors in all microscopic regions are also aligned when the solution is forced through a spinneret, to create a fibre when the sulphuric acid is removed. So the liquid crystallinity which exists as the fibre is spun from the solution ensures that essentially all of the chains in the resulting fibre are aligned with their long axes down its length. Typically, commercial

Kevlar is 50–80 per cent crystalline, but even in the amorphous regions the chains are close to being aligned. In the crystalline regions the chains adopt a planar structure, in general terms not unlike that in crystalline polyethylene (Fig. 3.2). But in Kevlar, the chains cannot be folded (Fig. 4.10) and other defects and irregularities of the types common in polyethylene do not exist. The result is that the yellowish fibres (typically of around 12 μm diameter) are enormously strong: any tensile stress for instance can only be relieved by breaking chemical bonds when there are no disorganized chains to be realigned, as in say PET (Fig. 4.10). In general nylon 6,6 fibres are considered to be very strong, but they offer no contest to Kevlar as is illustrated in Fig. 9.5. Kevlar cable can have a greater tensile strength than a steel cable of the same diameter, the latter being some five times heavier.

The strength of Kevlar does not originate in the individual chain structure *per se*, a point which is borne out by theoretical predictions that a polyethylene fibre with perfect crystallinity would have a comparable tensile strength. But until some means is discovered of producing polyethylene fibres devoid of defects, this is of passing interest only.

Unfortunately Kevlar is too expensive for general usage; its main use is as cord strengthening in vehicle tyres. 'Bullet-proof' vests and flak jackets are made of woven Kevlar cloth, up to 20 layers thick in some cases. It is amazing that such lightweight garments have the ability to stop bullets, often even at point-blank range. At temperatures approaching the melting point (T_m) of nylon 6,6 (about 530 K), the tensile strength of Kevlar fibre remains high, being about the same as that of nylon 6,6 fibre at ambient temperatures. The useful strengths of aramid polymers at temperatures of 600 K or slightly more commends their application in making heat-resistant materials, such as the Nomex linings commonly found in the cockpits of

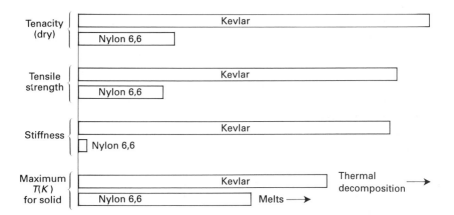

Fig. 9.5. Pictorial comparison of some of the properties of Kevlar and nylon 6,6 fibres.

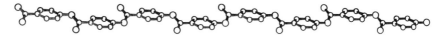

Fig. 9.6. Representation of the chain structure of poly(p-hydroxybenzoic acid). Reprinted with permission from Blackwell, J., Liesser, G., and Gutierrez, G. A. (1983). Structure of p-hydroxybenzoate/ethylene terephthalate copolyester fibres. *Macromolecules*, **16**, 1418–22. Copyright (1983) American Chemical Society.

racing cars: this material offers additional protection to the driver in the event of fire, since it burns only with difficulty and without melting, giving a thick char which creates a thermal barrier.

The first thermotropic liquid crystalline polymer was synthesized in 1976 and was effectively a copolymer based upon the constitutional repeating units (CRUs) of poly(p-hydroxybenzoic acid) (PHB) and poly(ethylene terephthalate) (PET). The homopolymer of p-hydroxybenzoic acid has the chain structure represented in Fig. 9.6. The rigid of chain of PHB itself results in the bulk material being a crystalline polymer which does not melt at temperatures below those at which decomposition becomes quite rapid, so that it is thoroughly intractable. It is only when another component is incorporated into the PHB chain that the rigidity eases enough for the bulk polymer to melt to give rise to a liquid crystalline mesophase. The substitution of at least one in five PHB residues by a CRU of PET is sufficient to give this phenomenon, in which the director orientates the rigid sections composed of linked PHB residues. The variation of the viscosity of the molten copolymers as a function of their composition provides evidence. At 548 K and starting with pure PET, the copolymers show a continuous increase in the viscosity of the melt as the proportion of p-hydroxybenzoic acid rises to 30 per cent on a molar basis. But further increase produces a dramatic fall in the melt viscosity coefficient, by over an order of magnitude in fact, to reach a minimum value at around 60 per cent. Thereafter, the melt becomes more viscous as the p-hydroxybenzoic acid content rises above 70 per cent. The range of 30–70 per cent results in the appearance of a nematic mesophase, in which the substantial degree of chain alignment is expected to reduce chain entanglement compared to a normal molten polymer. At the macroscopic level this is manifested in the very low viscosity coefficients in the range of existence of this thermotropic liquid crystal polymer.

The introduction of 'flexible spacers', such as the PET constitutional repeating units between rigid PHB segments in the case of the preceding paragraph, is one strategy for generating thermotropic liquid crystalline polymers (TLCPs). Another strategy is to disrupt the too intensive packing of linear chains by introducing a small extent of irregularity into them. Two TLCPs, polyesters in general nature, have the chain structures represented in Fig. 9.7. The trade names of these are Vectra and Xydar (or Ekonol)

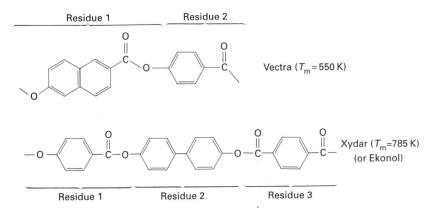

Fig. 9.7. Representations of the structures of the CRUs of the chains of Vectra and Xydar (Ekonol), showing the monomer residues.

respectively. The random copolymerization of two (Vectra) or three (Xydar) monomers of differing lengths and perhaps some degree of bentness (as in Vectra) allows TLCP behaviour to be observed above the melting point (T_m). In the TLCP condition, the viscosity of the molten polymer is very much lower than those of melts of common polymers, such as PET for example. Very low viscosity makes for easy and complete filling of moulds and TLCPs are thus favoured for making, say, electronic components, when complex shapes and thin walls must be precisely formed on the miniature scale. It is also advantageous in this that TLCPs solidify with almost no change in volume, which is essential for the production of precision components by moulding, be it on the large or small scale.

The examples given in this section are all main-chain liquid crystal polymers, the only type which has achieved main commercial significance at present. But side-chain liquid crystal polymers offer outstanding promise for the future and remarks will be made on these in Chapter 10 accordingly (Section 10.2).

9.3 High-temperature polymers

In Section 3.5, the argument was developed which related low values of the entropy of melting (ΔS_m) of a polymer to high values of T_m, the highest temperature at which any part of the solid phase exists. The implication is that stiff chains, at best only able to undergo internal rotation with difficulty, are the basic requirement of a polymer which has a value of T_m in the high-temperature range, say, above 600 K.

When the entire backbone of a polymer molecule is composed of two (or more) strands of linked atoms, it is said to have a *ladder* structure, using

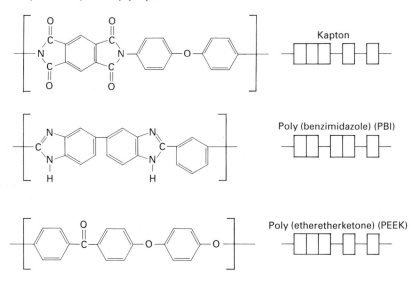

Fig. 9.8. Representations of the structures of the CRUs of the chains of Kapton, PBI, and PEEK (*left*) and the corresponding semiladder structures (*right*).

an evident analogy. A chain is said to have a *semiladder* structure if the backbone is composed of double-stranded sections interspersed with sections in which there is only one strand of linked atoms. The high-temperature polymers which have achieved some commercial significance are semiladder in type, based upon ring structures with imido units (>C−N<) or ether (−O−) linkages: examples are shown in Fig. 9.8. On the right-hand side, below the trade name of the polymer, is a simplified representation indicating the structure of the constitutional repeating unit, where boxes correspond to double-stranded parts of the backbone. The carbonyl (>C=O) group in PEEK is represented as a box because, as a component of a conjugated system of bonds, it is held in the plane of the adjacent aromatic rings. Kapton and PEEK have similar structures in terms of the representations on the right, but the polyimide, Kapton, is slightly more stable with an upper limiting temperature of usefulness of around 900 K, whilst PEEK is limited to temperatures below 800 K. PBI is roughly intermediate in this respect, but it can withstand short exposures to temperatures of around 970 K. All of these polymers have T_m values which are above their thresholds for fairly rapid thermal decomposition when air is present.

The polymers represented in Fig. 9.8 have significant degrees of crystallinity (particularly PEEK) but this is not always a characteristic of bulk polymers in which the chain is rather stiff (see Fig. 4.15 for example). It may seem odd at first sight that polymers with rather rigid chains may not have much crystallinity, or even none at all. Space does not permit a detailed discussion and it must suffice to say simply that there are certain structural features in some constitutional repeating units which prevent the individual chains packing closely together, the prerequisite for crystallinity. For instance there are certain substitutional patterns of phenyl rings which can force them out of coplanarity with one another, so preventing the chains packing together regularly even if they are rod-like and mutually aligned to a substantial extent.

A polyimide like Kapton is synthesized by a melt–condensation process which is often referred to as the dianhydride–diamine route. Figure 9.9 represents the two main stages of this synthesis when the starting materials are the anhydride of pyromellitic acid and 4,4′-diaminodiphenylether, which give rise to a polyamic acid in the first stage. In the subsequent stage the temperature is raised to around 575 K under elevated pressure (typically 15–20 times atmospheric pressure), to induce the ring closures with the evolution of water. In the second stage a key feature is the exclusive formation of the five-membered ring via an intramolecular mechanism: there is no formation of crosslinks between different chains through intermolecular condensation reactions. Fibres and films are in fact produced by mechanical working of the intermediate polyamic acid before its thermal dehydration to Kapton. This is simply because the polyamic acid

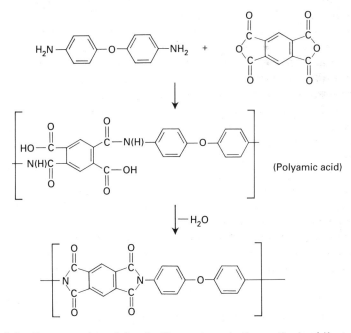

Fig. 9.9. Representation of the significant stages in the synthesis of Kapton.

is tractable, largely reflecting the facts that its chain has relatively little ladder structure and substituents are available which can enter into hydrogen bonding. As a result the polyamic acid is quite soluble in solvents such as dimethylformamide, contrasting with the insolubility of Kapton.

PBI polymers are synthesized by condensation of the diphenyl ester of isophthalic acid and a biphenyl tetramine, the residue of the latter being apparent in the central part of the PBI structure shown in Fig. 9.8. Isophthalic acid itself is not used as a monomer, because of the difficulty which would attend forming a melt solution of the monomers in the correct proportions at temperatures below 570 K, when the melting point of the pure acid is 621 K. In the first stage of the reaction, the prepolymer is formed via eliminations of water between $-NH_2$ and $C_6H_5-C(=O)$ $-O-$ groups to produce $-N=C(OC_6H_5)-$ links. This prepolymer is separated out as a solid and then taken up to temperatures around 100 K higher than those in the first stage: five-membered rings are induced to form via elimination of phenol (C_6H_5OH) between $-OC_6H_5$ groups in the linkage and $-NH_2$ groups in the vicinity. It is worth noting that PBI polymers are strongly coloured, ranging from yellow to almost black, which would be expected when the chains contain conjugated systems of bonds.

Polyimides are too expensive for general usage, but they are found in situations which require their abilities to survive exposure to high temperatures. They are very useful for making printed electronic circuits since it is possible to solder on the films concerned without melting or burning them. Several components of jet aircraft engines are commonly made of polyimide. As the material in compressor vanes they are advantageous, being relatively lightweight but robust in the hostile environment which precludes the usage of lubricants in general.

Poly(etheretherketone) (PEEK) is synthesized by reaction of the potassium salt of hydroquinone with difluorobenzophenone in the high-boiling solvent diphenylsulphone at temperatures close to the T_m value of the polymer (around 608 K). The process may be represented as

$$n K^+ O-C_6H_4-O^- K^- + nF-C_6H_4-C(=O)-C_6H_4-F$$

$$\xrightarrow{\text{550-610 K}} -[O-C_6H_4-O-C_6H_4-C(=O)-C_6H_4-]_n + 2n\ KF.$$

It is the use of the expensive fluorinated reactant which is largely responsible for PEEK being uneconomic for general usage. In bulk, PEEK is grey and opaque in appearance. In its speciality usages, its principal attractions are its thermoplastic processability coupled with the maintenance of high mechanical strength and resistance to decomposition up to relatively high temperatures.

9.4 Polycarbonate

Around 99 per cent of the world production of commercial polycarbonate is still the original material, bisphenol A polycarbonate, which was introduced to the market in 1959. It has the advantage of being prepared ultimately from acetone and phenol, chemicals which are inexpensive and readily available. The other key reagent is phosgene ($COCl_2$) and the overall process may be represented as

$$2C_6H_5-OH + CH_3-C(=O)-CH_3 \xrightarrow{\text{acid}} HO-C_6H_4-C(CH_3)_2-C_6H_4-OH + H_2O$$
$$(\text{Bisphenol A})$$

n Bisphenol A + n $COCl_2$ + $2n$ NaOH

$$\rightarrow -[O-C_6H_4-C(CH_3)_2-C_6H_4-O-C(=O)-]_n + 2n\text{NaCl} + n\text{H}_2\text{O}.$$

The synthesis is often conducted in the two-phase system created by the immiscible liquids dichloromethane (CH_2Cl_2) and water. Phosgene gas is bubbled into the liquids which are vigorously agitated to ensure that the organic reactants in the CH_2Cl_2 phase achieve good contact with the sodium hydroxide (NaOH) in the aqueous phase. The process is conducted at temperatures not far from ambient and cooling is required since the process is quite strongly exothermic. Both batch and flow (continuous) reactor systems are used widely. The typical result is the production of 15–20 per cent w/w of polycarbonate in CH_2Cl_2 initially, which gives rise to the amorphous polymer on removal of the solvent by evaporation. The industrial processes are specifically designed to yield polycarbonate in the amorphous form, which has the desired properties of extreme toughness, optical clarity, and high impact strength, the last being preserved over the wide temperature range of 75–410 K.

Outstanding features of polycarbonate are that the bulk amorphous material includes a free volume which is around 1½ times larger than that for most other amorphous polymers and it is resilient towards impacts at temperatures well below the glass transition temperature ($T_g \approx 432\,\text{K}$). There is no full explanation which is generally accepted, but it seems likely that two types of motion can occur at the microscopic level. A so-called α relaxation process depends on normal segmental rotation associated with the glass transition (Section 3.4). But there is a second (β) relaxation process which depends upon some lesser form of facile motion within the assembly of chains, which offers a means of dissipating localized stress. It seems likely that the motion associated with the β process is probably located on the backbones of the polymer chains and is sufficient both to give rise to the relatively large free volume and to allow the chains to avoid rupture under impact.

There are straightforward reasons for bisphenol A polycarbonate being the only commercially significant polymer of this type. In the first place it is relatively inexpensive, allowing its wider use under trade names such as Lexan. Then it can serve as unbreakable glass: the chain has a relatively low symmetry which promotes the amorphous phase and makes the material transparent in bulk (Section 3.3) whilst impacts on it are dissipated by the β process. It becomes plastic at temperatures around 490 K, allowing it to be injection- or blow-moulded into complex shapes if these are required, or hot rolled into sheets. The material is also thermally stable (see Exercise 9.8) and is able to withstand a temperature of 600 K for a short time. It appears to be the aromatic rings ($-C_6H_4-$) in the chains which are essential for this resistance to thermal decomposition: the equivalent chains with alkyl linkages in place of these aromatic rings are much less stable, probably because they have β hydrogen atoms which are relatively easier to detach than any of the hydrogen atoms in bisphenol A polycarbonate.

9.5 Ionomers and polymeric solid electrolytes

These polymers have the common feature of containing ions or charge centres and are differentiated by the degree. *Ionomers* have a relatively low ionic content (<10–15 per cent on a molar basis), whilst *polyelectrolytes* have considerably larger densities of charges.

Ionomers have hydrophobic backbones to which pendant groups bearing charges are attached. Figure 4.12 refers to a copolymer bearing sulphonate ($-SO_3^-$) groups with either Na^+ or Zn^{2+} counterions in their vicinity: the SAXS profiles are interpreted in terms of the formation of the formation of domains. The hydrocarbon chain of the polymer and its salt-like substituents are obviously incompatible but the incompatibility is resolved when the charge centres agglomerate to form microregions of an ion-rich phase (i.e. domains) embedded in the nonpolar matrix of the rest of the polymer chains. Figure 9.10 shows photographs of small areas of sulphonated polystyrenes (Zn^{2+} counterions) taken using an electron microscope, which reveal domains about 3–4 nm across.

The commonest ionomers are copolymers of 1-alkenes with methacrylic acid ($CH_2{=}C(CH_3)$ (COOH)), produced when the carboxylic acid group is partially neutralized to give a carboxylate group ($-COO^-$) with a counterion, usually Na^+, Mg^{2+} or Zn^{2+}, in its vicinity. Figure 9.11 shows possible models of the structures of the ion-rich microregions which form in such bulk ionomers.

The domains in ionomers represent points at which different chains are anchored together and so act in the same way as crosslinks, such as those in vulcanization (Fig. 3.4). But the domains break up at high enough temperatures, up to 475 K in some cases, so that they function really as physical rather than chemical crosslinks. Segmental rotations in the matrix of the ionomer are rapid, when the temperature is above the corresponding

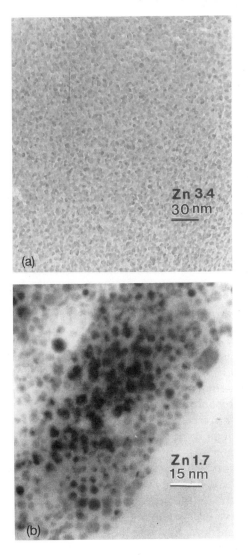

Fig. 9.10. High-voltage (transmission) electron microscope (HVEM) photographs of extremely thin films of Zn^{2+}-neutralized sulphonated polystyrenes at critical focus (a) 3.4 mol % sulphonation (b) 1.7 mol % sulphonation (the featureless region in the lower right-hand corner is beyond the edge of the film). Reproduced from Li, C., Register, R. A., and Cooper, S. L. (1990). Direct observation of ionic aggregates in sulphonated polystyrene ionomers. *Polymer*, **30**, 1227–33. by permission of the publishers, Butterworth Heinemann Ltd.

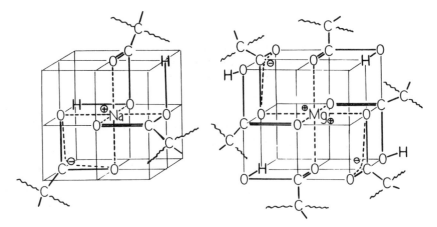

Fig. 9.11. Representations of the ordered structures of ionic domains within copolymers of ethylene and methacrylic acid which have been neutralized by sodium (Na^+) and magnesium (Mg^{2+}) ions. Reprinted by permission of John Wiley & Sons, Inc., from Hirasawa, E., Yamamoto, Y., Tadano, K., and Yano, S. (1991). Effect of metal cation type on the structure and properties of ethylene ionomers. *Journal of Applied Polymer Science*, **42**, 351–62. Copyright (1991) John Wiley & Sons, Inc.

glass transition temperature. This means that ionomers have the general characteristics of thermoplastic elastomers, simply having a different type of physical crosslinking to those discussed in Section 8.4. Ionomers have much better mechanical properties than the parent (co)polymer, being considerably stronger and in particular stiffer. Just like thermoplastic elastomers, they can be injection moulded to produce desired shapes, and they are attractive in other respects, such as having good resistance to abrasion and to oil whilst being reasonably flexible. But perhaps the most surprising feature is their optical transparency, which is attributed to the entirely amorphous structure of the matrix of the ionomer, with the embedded ion-rich domains being too small to scatter light significantly. It appears that the ionic domains disrupt any tendency for crystallites to form within the matrix.

An ionomer which has achieved some significance commercially is a copolymer of ethylene and methacrylic acid containing sodium or zinc cations, commonly referred to by the trade name Surlyn. The copolymeric chain is synthesized by a radical addition mechanism under high pressure conditions and the monomer ratios are adjusted to give about 5 per cent of acidic residues in the product. This is converted into the ionomer by melting, in-mixing of a suitable salt, often sodium carbonate, magnesium hydroxide or zinc oxide, and allowing reaction to occur in the melt. The salts used in this connection have an anion which will be destroyed and effectively taken off as a volatile product, usually carbon dioxide or water.

Thereafter the ionomer is extruded, cooled, and cut into pellets for ease of handling.

The creation of a polymeric solid electrolyte was first demonstrated some 20 years ago. Major interest has centred on the class of these materials known as *solvent-free polymer ionics*, of which the commonest cases are poly(ethylene oxide) (PEO) or copolymers of ethylene oxide and propylene oxide when these contain rather large concentrations $(0.5-3 \, \text{mol dm}^{-3}$ typically), of an alkali metal salt, denoted generally as MX. Common identities of M^+ are Li^+ or Na^+, whilst X^- is often perchlorate (ClO_4^-). The common method of preparation is to dissolve the polymer and MX in a suitable solvent, water when PEO is the polymer, which is then evaporated off to leave the product as a bulk sample or as a thin film, depending upon the application for which it is required. One indication of the profound difference between the starting polymer and the solvent-free polymer ionic material derived from it is a considerable rise (generally by at least 60 K) in the glass transition temperature resulting from the conversion. Confirmation that the product is not just simply a physical mixture is provided by its different infrared absorption spectrum compared to the initial polymer. But of course the outstanding difference is that the new material conducts electrical current whilst the starting polymer is an insulator.

The alkali metal cations (M^+) in polymeric solid electrolytes are considered to be bound coordinatively to Lewis base sites (i.e. sites capable of donating a pair of electrons with relative ease) which are distributed along the polymeric chains. For instance, in PEO the CRUs are joined by ether linkages $(-CH_2-O-CH_2-)$ in which the oxygen atoms have pairs of electrons which are not involved in chemical bonding and are thus available to be donated to M^+. The resultant key elements in this structure are shown in Fig. 9.12(a). It is generally accepted that it is the continuous amorphous phase which provides the electrical conduction mechanism, even if these solvent-free polymer ionic materials created from PEO tend to be highly crystalline. A key point is that the glass transition temperature (T_g) must be well below the temperature of usage as an electrical conductor, because it is segmental rotation which is crucial for the mechanism of charge transport. Figure 9.12(b) represents the essential dynamical features: the initial chain conformation (i) is converted into the conformation (ii) by rotation about the backbone bonds indicated by the two shorter arrows. The anion (X^-) must follow the positive charge and so moves as indicated by the large arrow. Thereafter it is to be imagined that this X^- moves further to the right under the influence of the electrical field and becomes associated with a different positive charge on another chain to the right of that shown; at the same time another X^- anion enters on the left. A further segmental rotation of the chain shown can be imagined to bring it back to conformation (i) when the M^+ bound

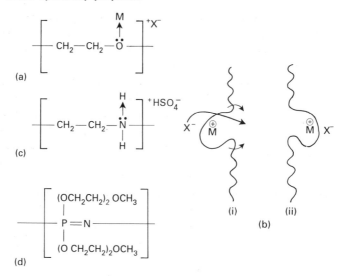

Fig. 9.12. Representations of significant microscopic structures in polymeric solid electrolytes (see text for details).

to it is brought back to make contact with the 'new' X^- anion. Repeated actions of this sort result in the movement of the anions from the left (i.e. from the negative electrode) to the right (towards the positive electrode), thus carrying electrical charge. The cations, M^+ cannot be effective charge carriers when they are anchored firmly to polymer chains. This is the simplest model which has the anions moving through the continuous amorphous phase under the influence of the applied electrical field. The mean separation of the ions is typically only around 0.5–1.0 nm, reflecting the large concentrations of MX incorporated, so that each ion will be affected by the coulombic field created by neighbouring charges.

Organic polymer electrolytes have good electrical conductivity when they are almost completely amorphous. Crystalline zones obviously do not permit the conduction mechanism to operate, so that partially crystalline materials are invariably less conductive. Poly(ethylene imine) (PEI) becomes conductive when concentrated sulphuric acid is incorporated, the effective structure being represented in Fig. 9.12(c). The architecture of the polymeric chain can have a spectacular effect on the resulting electrical conductivity. Branched PEI is completely amorphous and with a sulphuric acid content corresponding to 0.35 of an H_2SO_4 molecule per monomer residue, it shows a conductivity of $8.5 \times 10^{-3}\,S\,cm^{-1}$ at 298 K. On the other hand, PEI with an unbranched (i.e. linear) chain is highly crystalline, as would be expected on the basis that its chains will pack readily together (in contrast to branched chains). With the same sulphuric acid content as the branched PEI specified above, linear PEI only has an electrical conductivity of the order of $10^{-8}\,\mathrm{S\,cm^{-1}}$. This dramatic difference is a further demonstration of the importance of chain architecture for the properties of bulk polymers.

Some recent research efforts in connection with solid polymer electrolytes has focused on comb-like chains (Section 9.1) which have very flexible backbones, the latter discouraging crystallinity. Examples are polysiloxanes (Fig. 3.7) and polyphosphazenes, bearing side chains which correspond to small oligomers of ethylene oxide. The CRU of a polyphosphazene chain of this type is shown in Fig. 9.12(d). The conduction arises when the salt MX is incorporated, when the oxygen atoms in the side chains form coordinative bonds with M^+ and X^- acts as the charge carrier.

Exercises

9.1. $\bar{M}_n$ values of commercial Kevlar samples are typically 1×10^4. What is the corresponding average number of CRUs per chain?

9.2. Below is a list of the T_g values for polymers with various groups R in the CRU represented as $-CH_2-CH(R)-$

R	$-(CH_2)_2CH_3$	$-CH(CH_3)_2$	$-(CH_2)_3CH_3$	$-CH_2CH(CH_3)_2$
$T_g(K)$	233	323	223	302
R	$-(CH_2)_3CH_3$	$-(CH_2)_2CH(CH_3)_2$	$-C(CH_3)_3$	$-CH_2C(CH_3)_3$
$T_g(K)$	242	259	337	332

What trends may be discerned and how may these be interpreted?

9.3. Poly(p-phenylene terephthalate) which has the CRU structure $-C_6H_4-O-C(=O)-C_6H_4-C(=O)-O-$ shows thermotropic liquid crystal behaviour in which the solid becomes a nematic phase on rising through around 650 K, whereas poly(ethylene terephthalate) shows no liquid crystal behaviour. How may this be explained? What behaviour would be expected of a random copolymer of the two CRUs concerned in the proportions of phenylene:ethylene 9:1?

9.4. Using the description given in Section 9.3, draw structural formulae which indicate how PBI is synthesized from its monomers (see Fig. 9.8).

9.5. A very strong fibre is produced when the polyimide having the CRU structure represented as

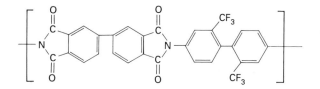

has been spun from its isotropic solution (lyotropic liquid crystalline state) in m-cresol and subsequently drawn in air. A fibre resulting from application of a draw ratio (λ, see Section 3.2) of 8 has been investigated using wide-angle X-ray scattering (WAXS, Section 4.3.1) to reveal a unit cell of total volume 2.559 nm^3 which contained four CRUs. Degrees of crystallinity (X) were obtained by WAXS measurements and corresponding densities (ρ) by a density column method. The following results were obtained

Sample description	$X(\%)$	ρ (kg dm^{-3})
As-spun fibre	9.8	1.446
Drawn fibre ($\lambda = 8$)	51.1	1.471

(i) What is the relative molecular mass of the CRU?
(ii) What is value of ρ within perfectly crystalline regions?
(iii) What is the value of ρ for the fully amorphous polyimide?
(iv) Comment on the values of ρ of the crystalline and amorphous phases in comparison with those listed in Table 4.2.

Source: Cheng, S. Z. D., Wu, Z., Eashoo, M., Hsu, S. L. C., and Harris, F. W. (1991). *Polymer*, **32**, 1803–10.

9.6. Polyimides of the general type represented by the structure shown in Fig. 9.13 have been synthesized. The enthalpies (ΔH_m) and entropies (ΔS_m) for the melting of the crystallites have been measured for n values as follows

n	1	2	3
ΔH_m (kJ mol^{-1})	72.5	80.2	88.0
ΔS_m (J $mol^{-1} K^{-1}$)	118.3	139.0	162.7

(mol refers to the CRU structure ($m = 1$) given below.)
(i) Evaluate T_m for each polyimide.
(ii) Compare the data above with entries in Table 3.3 (in particular that of poly(ethylene oxide)) and comment.

Source: Cheng, S. Z. D., Heberer, D. P., Lien, H-S. and Harris, F. W. (1990). *Journal of Polymer Science*, **B28**, 655–74.

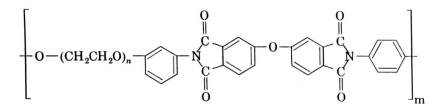

Fig. 9.13. General structure of the polyimides referred to in Exercise 9.6.

9.7. A thermoplastic elastomer has been synthesized by step-growth polymerization commencing with a poly(ethylene oxide glycol) ($H-(OCH_2CH_2)_n-OH$) of $\bar{M}_n = 1000$, dimethyl terephthalate ($CH_3O-C(=O)-C_6H_4-C(=O)-OCH_3$), and 1,4-butanediol ($HO-(CH_2)_4-OH$). The resulting poly(ether ester) with $\bar{M}_n = 2708$ gave rise to the traces shown in Fig. 9.14 in differential scanning calorimetric (DSC) studies (Section 3.6) on the

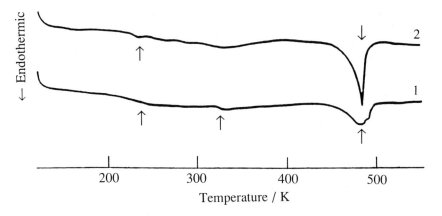

Fig. 9.14. Differential scanning calorimetry (DSC) traces of the poly(ether ester) referred to in Exercise 9.6. Reproduced by permission of Akademie Verlag GmbH, Berlin, from Gogeva, T., Fakirov, S., Mishinev, J., and Sarkisova, L. (1990). Poly(ether ester) fibres. *Acta Polymerica*, **41**, 31–6.

as-synthesized polymer (lower trace) and the drawn fibre (draw ratio of 3.64) (upper trace).

(i) What is the average degree of polymerization of the poly(ethylene oxide glycol)?

(ii) Specify the composition of the average poly(ether ester) chain in terms of residues.

(iii) Discuss how the features indicated by the arrows in Fig. 9.14 may be interpreted.

(iv) What microscopic features are responsible for the thermoplastic elastomeric nature of the polymer?

9.8. Bisphenol A polycarbonate decomposes thermally in the absence of air at 773 K. Sampling of the evolved products using a mass spectrometer revealed the presence of prominent species with mass numbers 44, 94, 107, 120, 122, 136, 228, and 508. What are the most likely identities of the molecules corresponding to these mass numbers?

Source: McNeill, I. C. and Rincon, A. (1991). *Polymer Degradation and Stability*, **31**, 163–80.

9.9. A model of the ionomeric system created by dissolution of sodium iodide (NaI) in polyethylene oxide (PEO) has each Na^+I^- ion pair complexed by three consecutive monomer residues in a PEO chain. Given that the lattice energy of pure sodium iodide is $-704\ kJ\ mol^{-1}$, that its lattice energy within the complex is $-540\ kJ\ mol^{-1}$ and that in crystalline PEO each chain has a binding energy equivalent to $-13.2\ kJ$ per mole of monomer residues, evaluate the least negative binding energy of the complex which is needed to explain the fact that sodium iodide dissolves spontaneously into the PEO polymer. Assume that there is no mutual interaction between the PEO chains in the ionomer.

Source: Wright, P. V. (1989). *Polymer*, **30**, 1179–83.

9.10. Bitumens can be regarded as being mainly polyalkenes; their mechanical properties are considerably improved in blends with the triblock copolymer styrene–butadiene–styrene (SBS). But simple blends are not thermodynamically stable so that demixing is a deterrent to practical usage. Bitumen and SBS copolymer were both chemically modified by grafting on carboxylic acid groups before they were mixed. Zinc acetate was incorporated into the mixture at 443 K, when acetic acid was evolved. The resultant blend did not undergo demixing over long periods of time. How may this be explained?
Source: Engel, R., Vidal, A., Papirer, E., and Grosmangin, J. (1991). *Journal of Applied Polymer Science*, **43**, 227–236.

10
Looking to the future

This relatively brief chapter is aimed at indicating some of the interesting developments which are likely to become important in the near future in connection with synthetic polymers. Methods which are known but little exploited at present, for synthesizing polymers with complex molecular structures are outlined before some prospective new uses of polymeric materials are discussed.

10.1   The synthesis of block copolymers

Tailored polymers and in particular block copolymers, which have been synthesized to achieve a particular chain size and architecture, seem likely to become increasingly important in the future. Thus the major interests in this section are copolymers of specified types, particularly those of diblock and triblock natures which are synthesized from monomers which are not able to be polymerized by anionic mechanisms (Section 7.3).

Iniferters appear to offer very considerable potential: the term iniferter is an abbreviated form of *ini*tiation-trans*fer*-agent *ter*minator. The systems of main interest can be regarded as 'living radical' polymerizations, when the iniferter, designated as $I-I$, can be described as an initiator which is also an effective chain transfer agent, (Section 6.5). Thus the action of the iniferter in the presence of a suitable monomer (M) can be represented as

$$I-I \qquad \rightarrow \qquad I\cdot + I\cdot$$

$$I\cdot + nM \qquad \rightarrow\rightarrow\rightarrow \qquad IM_n\cdot$$

$$IM_n\cdot + I-I \qquad \rightarrow \qquad IM_nI + I\cdot$$

The evident result is that a polymeric block (M_n) has been inserted into the bond of the original $I-I$ molecule. The systems of real interest are those in which $I-I$ is a functionalized polymeric molecule of a different type from M_n.

The potential usage of iniferters may be understood best through consideration of an actual example. Thiuram sulphide molecules have as a central feature the linkage represented as $>N-C(=S)-S-S-C(=S)$ $-N<$, within which the $S-S$ bond is subject to thermal dissociation. In the example chosen, the initial reactants were in turn various oligomeric dimethylsiloxanes (designated as ODMS) with average degrees of polymerization within the range 5–42 and suitably functionalized for conversion to the corresponding chains composed of various numbers of units represented as $-ODMS-N(CH_3)C(S)S-SC(S)N(CH_3)-$, in processes which used carbon disulphide and iodine as the principal reagents. These products may be described as oligomeric iniferters and they are active in this respect at temperatures of 350 K or so. When the thiuram sulphide S—S bonds of this iniferter break in the presence of methylmethacrylate (MMA), this monomer is incorporated as blocks of MMA residues, so that the result is a multiblock copolymer composed of repeating units of the general type $-[ODMS-N(CH_3)C(S)S-[CH_2-C(CH_3)(COOCH_3)-]_nSC(S)N(CH_3)-]$.

In themselves, polysiloxanes have several useful properties, such as impact and oxidative resistance combined with easy processability, but these are offset by mechanical weakness. But when poly(methylmethacrylate) blocks have been incorporated with siloxane blocks in the backbone, the resultant multiblock copolymer is considerably stiffer and stronger than the corresponding polysiloxane, whilst preserving most of the desirable properties of the latter. When these multiblock copolymers were investigated by differential scanning calorimetry (DSC) (Section 3.6), two glass transitions were revealed. One of these corresponded to $T_g \approx 140\,K$, which is close to that observed in polydimethylsiloxane (see discussion of Fig. 3.7) and is thus ascribed to loosening of internal rotations in the ODMS blocks. The second glass transition corresponded to $T_g \approx 405\,K$, close to that of poly(methylmethacrylate) (Table 3.1); evidently above ~405 K, there is rapid internal rotation within the poly(methylmethacrylate) blocks. But at ambient temperature the rigidity of the poly(methylmethacrylate) blocks reinforces the otherwise mobile ODMS blocks, conferring mechanical strength on the bulk copolymer. As with most block copolymers of commercial interest, the different types of monomer residues are incompatible and as a result form separate domains in the bulk polymer, as was seen in Fig. 8.8 for another block copolymer.

A second approach to the synthesis of block copolymers has become known as the *macromonomer* technique, reflecting its usage of *macromolecular monomers*. Basically this technique involves the initial synthesis of oligomeric chains via one type of mechanism, before these oligomers are applied as if they were monomers in another type of polymerization system. Again this is developed best through an actual example. Initially methylmethacrylate (MMA) has been oligomerized using AIBN as the initiator and thiomalic acid $(HOOC-CH(SH)-CH_2-COOH)$ as the chain transfer agent in a radical addition system (Section 6.5) The growing chains in this case are terminated predominantly by the chain transfer reaction represented as

$$----\cdot + HOOC-CH-CH_2-COOH \rightarrow ----H + HOOC-CH-CH_2-COOH.$$
$$\quad\quad\quad\quad\quad | \quad\quad\quad\quad\quad\quad\quad\quad\quad\quad\quad\quad\quad\quad\quad |$$
$$\quad\quad\quad\quad\quad SH \quad\quad\quad\quad\quad\quad\quad\quad\quad\quad\quad\quad\quad S\cdot$$

As the resulting radical on the right initiates new chains, it confers on them the dicarboxylic acid functionality which is used in the following stage when these macromonomers are used in a step-growth polymerization with a diamine, such as hexamethylene diamine (similar to nylon 6,6 synthesis (Fig. 1.8)). The resulting polyamide can be represented as

$$-[N(H)-(CH_2)_6-N(H)-C(=O)-CH-CH_2C(=O)-]_n.$$
$$\quad\quad\quad\quad\quad\quad\quad\quad\quad\quad\quad\quad\quad\quad |$$
$$\quad\quad\quad\quad\quad\quad\quad\quad\quad\quad\quad\quad\quad\quad S$$
$$\quad\quad\quad\quad\quad\quad\quad\quad\quad\quad\quad\quad\quad\quad |$$
$$\quad\quad\quad\quad\quad\quad\quad\quad\quad\quad\quad\quad (MMA)_m$$

This may be regarded as a graft copolymer (Section 9.1), with blocks of MMA residues pendant at regular intervals along a polyamide chain.

Macromonomers are more commonly synthesized by living ionic polymerization (Section 7.3), by deliberately 'end-capping' the living end with a residue possessing a vinyl-type C=C bond; this is then a macromonomer which can be used in a radical addition system with a different monomer from that in the initial ionic system. For example anionic living polystyrene chains with Li^+ gegenion may be terminated by reacting with allyl bromide, this action being represented as

$$-----\ominus Li^+ + BrCH_2CH=CH_2 \rightarrow -----CH_2CH=CH_2 + LiBr.$$

The macromonomers ($----CH_2CH=CH_2$) could then be used in a radical addition system, say with methylmethacrylate as the monomer, to produce multiblock copolymers with blocks of styrene residues attached to methylmethacrylate residues.

'Star' polymers have been discussed to some extent in Section 8.4, where interest was restricted to those with a few arms centred on a single residues of the molecule used as the coupling agent. Here the view is widened to consider *multi-armed star* polymers. All of the principal routes for synthesis of these macromolecules are anionic addition polymerizations and Fig. 10.1 represents the most general of these.

The upper mechanism in this representation shows a living polymer chain (say polystyrene (PS)) with K^+ as its gegenion initiating polymerization of a small amount of divinylbenzene (DVB) added to the system after consumption of styrene. DVB is a bifunctional monomer via its two CH=CH$_2$ groups (represented here simply as =). Complete consumption of DVB results in the formation of small nodules of poly(DVB), each of which is surrounded by the polystyrene arms from the original living chains

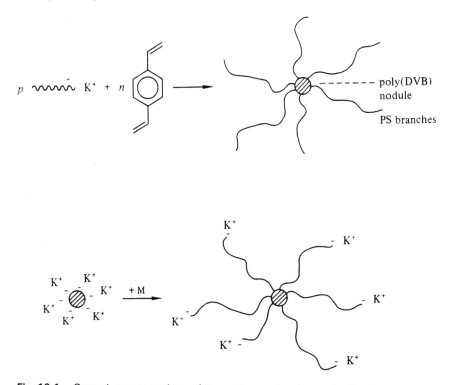

Fig. 10.1. General representations of two schemes for the synthesis of star-shaped macromolecules which are copolymers of polystyrene (PS) and poly(divinyl benzene) (poly(DVB)) (see text for details). Reproduced with permission of Hüthig & Wepf Verlag, Basel, from Lutz, P. and Rempp, P. (1988). New developments in star polymer synthesis. Star-shaped polystrenes and star-block copolymers. *Makromolekulare Chemie*, **189**, 1051–60.

which added onto any of the DVB molecules which are incorporated as residues in the nodule. The average number of arms per molecule (or nodule) can be varied by changing the ratio of the number of living chains to the number of DVB molecules. Typical polymers of this type have a poly(DVB) content of 1–5 per cent by weight and have $\bar{M}_n$ values of the order of 10^6. Potassium/naphthalene is the usual initiating system (Fig. 7.3) used to generate the living polystyrene chains.

The lower mechanism in Fig. 10.1 is a so-called 'core first' route. Microscopic particles bearing living anionic sites and balancing gegenions (K^+ say) are created by direct initiation of DVB polymerization in its dilute solution. Then the main monomer, say styrene, is added and chains grow outwards from each anionic site. The average number of arms per resultant star-polymer molecule is typically in the range 5–50. One further aspect of this route is that it offers the possibility of growing *star-block copolymers*, simply by allowing one monomer to be consumed and then adding a different one, before finally 'killing' the polymer. In one instance the inner blocks on the arms were polystyrene, whilst the outer blocks were poly(ethylene oxide). When the growth of these star-block copolymers was killed with addition of water, the arms were all tipped with hydroxyl (OH) groups. This seems to have the potential to be a useful polymer in various applications, such as action as a solubilizing agent for inorganic compounds in organic liquids.

10.2 Side-chain liquid crystals

The basis of liquid crystallinity has been described in Section 9.2, where the interest was confined to polymers with main chains which were responsible for liquid crystalline behaviour. Here the interest extends to those polymers which show liquid crystallinity as a result of possessing rather rigid side chains. Figure 10.2 illustrates in a simplified manner the

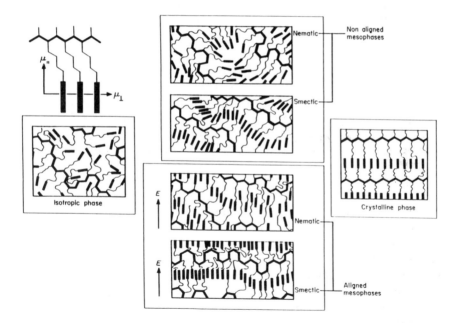

Fig. 10.2. Schematic illustrations of the chain structures and the organization of the chains in the different phases and mesophases of a liquid crystalline side-chain polymer. The mesogenic head group (filled rectangle) has longtitudinal ($\mu_{\parallel}$) and transverse ($\mu_{\perp}$) components of its dipole moment. E is a directing electric field. Reproduced from Attard, G. S. and Williams, G. (1986). The effect of a directing electric field on the physical structure of a liquid crystalline side chain polymer: the preparation of aligned thin films and their study using dielectric relaxation spectroscopy. *Polymer Communication*, **27**, 2–6, by permission of the publishers, Butterworth Heinemann Ltd. Copyright 1986.

types of phases and mesophases which can appear in such systems. The rigid parts of the side chains are represented as filled rectangles whilst the flexible main chains are represented as a heavy continuous lines. The lower parts of the central section represent the mesophases which can arise when an electric field (E) is applied to align the rigid parts of the side chains. It is evident that the main chains are relatively disordered in these mesophases in comparison with the crystalline phase on the right.

Side-chain liquid crystal polymers have clearly rather special regular types of comb structures, in which the 'teeth' are composed of a flexible 'spacer' section (the twisting thin lines in Fig. 10.2) connecting from the main chain to the rigid section of the side chain. As might be anticipated from the discussion of the preceding section, the commonest method of synthesizing these types of polymers centres on the free radical addition of monomers with a terminal $C{=}C$ bond joined to a flexible spacer group linking through to the rigid (or mesogenic) group. A representative monomer in this connection is $CH_2{=}C(CH_3){-}C({=}O){-}O{-}(CH_2)_5{-}O{-}C_6H_4{-}C({=}O){-}$ $O{-}C_6H_4{-}C({=}O){-}O{-}CH_3$, with the aromatic rings ($-C_6H_4-$) having the bonds shown in the *para* positions (1,4-linkage); this molecule is a methacrylate in general type. The $-(CH_2)_5-$ section corresponds to the flexible spacer, whilst the aromatic rings are the main components of the mesogenic group or simply mesogen. The commonest general structure of the mesogen is $-R_1{-}C_6H_4{-}L{-}C_6H_4{-}R_2-$, in which R_1 and R_2 may be groups such as $-CH_2{-}O-$ or $-CH_2{-}C({=}O){-}O-$. Typical linking groups (L) are $-CH{=}N-$ (Schiff's base), $-O{-}C({=}O)-$ (ester), and $-C({=}O){-}N(H)-$ (amide).

Side-chain liquid crystal polymers (LCPs) are not unlike normal liquid crystals composed of small molecules in several respects. But there are particularly noticeable differences in connection with films, in particular the opaqueness of those formed of side-chain LCPs. This is because directors in various parts of the film are randomly orientated, but the mesogens are more aligned within microscopic areas referred to as domains (see upper parts in the centre of Fig. 10.2). The boundaries between the multitude of domains scatter light, so creating the opacity (see atactic PMMA (Section 3.3)) Now when an electric field is applied the film formed from the side-chain LCP becomes transparent, simply because domains do not exist when there is only one director throughout the material (lower parts in the centre of Fig. 10.2). One of the more interesting potential applications for these side-chain LCP materials arises from the fact that the side-chain mesogens remain aligned when the electric field is switched off. Heating is required to provide enough energy for the separation of aligned side-chains, which is the prerequisite for the reformation of domains and hence the reappearance of opacity. Laser drawing and recording is thus possible on these transparent films: the resolution is good because the beam can be focused down to a spot of around 10 μm diameter, only within which it heats and renders opaque the parts of the film exposed to it. High resolution 'laser writing' has been demonstrated at ambient temperatures using smectic (type A) polysiloxanes, when speeds of movement of the laser spot of the order of $1 \, \mathrm{m \, s^{-1}}$ have been found to produce good results. This allows a real prospect that side-chain LCPs may be used to create compact discs which can not only be used for storing large amounts of information but will be erasable and thus reusable for storing different information. Already compact discs have been created in laboratories which have been subjected to tens of thousands of record/erase cycles, yet show no evidence of significant deterioration of performance as a consequence. Current research effort is directed at finding stable substances which can be incorporated into the side-chain LCP so that the composite will absorb the infrared laser light more efficiently, to allow recording speeds to be increased.

Other exciting prospects for the application of side-chain LCPs are in the field of optics, particularly in connection with telecommunication systems using optical fibres and in computers based upon the transmission of light rather than electrical currents. These are beyond the scope of this book, but the interested reader may follow up some of the relevant reading suggestions given in the Further Reading section.

10.3 Synthetic metals

The overwhelming majority of bulk polymers are electrical insulators. Around 20 years ago, the first polymeric materials were created which showed high enough electrical conductivity ($> 0.1 \mathrm{S \, cm^{-1}}$) to be described as electrical conductors. The bulk polymers concerned only became conducting after incorporation of some 'dopant', which was an inorganic electron acceptor, such as arsenic pentafluoride ($\mathrm{AsF_5}$) or iodine($\mathrm{I_2}$). Figure 10.3 represents a scale of electrical conductivity and located on it are copper, the actual metal usually used for electrical wiring, and some of the 'synthetic metals', as the polymeric conductors are sometimes known.

Polyacetylene with the *trans* configuration has a characteristic blue metallic sheen, but is only a weak semiconductor in the pure bulk form. The incorporation into it of $\mathrm{AsF_5}$ is believed to give rise to charge-transfer complexes at various points along the chains: one of these may be represented in simple terms as shown below

$$-(CH{=}CH)_n{-}CH{=}CH{-} + AsF_5 \rightarrow -(CH{=}CH)_n^+{-}\overset{\overset{\displaystyle AsF_5^-}{\uparrow}}{CH}{-}CH{-}.$$

The strongly electron withdrawing nature of the $\mathrm{AsF_5}$ molecule is considered to withdraw electronic charge from the conjugated chain. The resultant positive charge is stabilized by its delocalization over a section

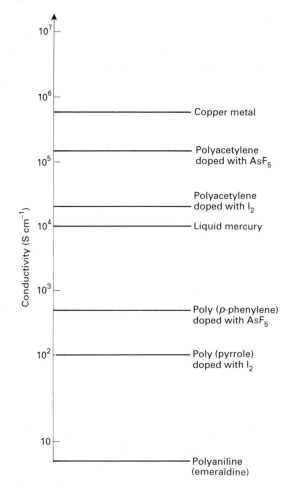

Fig. 10.3. Logarithmic conductivity ladder locating some metals and 'synthetic metals'.

$-(CH=CH)_n-$ with $n \approx 14$. In effect this positive charge can be viewed as a mobile 'positive hole' or polaron, which moves and thus transfers electrical charge (i.e. creates electrical current) under the influence of a potential difference. It is a similar ability to delocalize the unpaired electrons corresponding to polarons through conjugated systems of chemical bonds which is the general feature of the other polymer-based systems which appear in Fig. 10.3.

Another interesting phenomenon is that the electrical conductivity of doped polyacetylene increases when the material is stretched. Evidently the overall effect will be akin to drawing (Section 3.2, Fig. 4.9) and will be expected to result in greater degrees of alignment of the chains in the direction of stretching. Polarons can only travel through the bulk material if they can transfer from one polymer chain to another: it is believed that the individual acts of polaron transfer are particularly influenced by the extent to which adjacent sections of chains are aligned, being easiest when they are parallel. On this basis the increase in conductivity with stretching is expected.

Doped polyacetylene offers a particularly high electrical conductivity per unit density of material, in fact around three times that of copper metal. Thus it has attracted considerable interest with regard to its potential usage as electrical wiring or as electrode material in batteries. Unfortunately all of these 'synthetic metals' tend to be expensive in comparison with copper and suffer further disadvantages from difficulties of processing and ageing behaviour. In particular, aside from cost, it is the problems associated with general insolubility, infusibility, and brittleness which have inhibited wider application of these polymeric conductors. The last of these problems has been overcome with polypyrrole for example, by supporting a thin film of it on another film of a common plastic such as PET or poly(methylmeth-acrylate). The resultant composite combines the good electrical properties of the doped polypyrrole with the mechanical strength of the thermoplastic, giving a thin but robust conductor acceptable for commercial applications.

With the availability of lightweight electrodes made of say polypyrrole film which can be made into a sandwich around a 'filling' of a thin film of a solid polymer electrolyte (Section 9.5), electrical storage devices (i.e. rechargeable batteries) can be envisaged which will be paper-thin but will have power/weight ratios exceeding those of conventional lead–acid batteries by an order of magnitude. And of course these 'all-polymer' devices will be entirely free of leakage problems which are hazards with conventional batteries. Moreover the synthesis of copolymers which have conducting chains crosslinked by nonconducting links has already been achieved, which may be regarded as a first step on the way to producing single polymer molecules which can function as self-contained electronic devices. In turn this might be expected to lead on to the production of computing hardware in which the individual electronic elements are on the smallest scale imaginable. Huge computing capacities could then be contained in tiny volumes.

10.4 Polymers with active surfaces

Polymers have two properties which make them of interest as supports for active moieties. They are insoluble in common solvents in general and in particular in water. Also polymer chains can be functionalized in the sense that groups that are potentially reactive can be chemically attached to them. These properties make polymers of major interest in connection with ion-exchange and catalytic processes.

Ion-exchange resins are commonly copolymers of styrene and divinyl-benzene (DVB), which are functionalized subsequently with sulphonate $(-SO_3^-)$ groups to yield cation-exchange resins or with groups such as amino $(-NH_2)$ for anion-exchange resins. The copolymer is crosslinked by the DVB residues (Fig. 8.4), thus making it highly resistant to attack by solvents or chemical reagents. The usual appearance is amber-coloured beads. Undoubtedly the major usage of ion-exchange resins is in the deionization and softening of water supplies. For example, water which contains sodium chloride can be deionized by mixed cation-and anion-exchange resins, working through consecutive processes

$$----SO_3H + Na^+(aq) \rightleftharpoons ----SO_3^-Na^+ + H^+(aq)$$

$$----NH_2 + H_2O + Cl^-(aq) \rightleftharpoons ----NH_3^+Cl^- + OH^-(aq)$$

$$H^+(aq) + OH^-(aq) \rightarrow H_2O.$$

When they become saturated, the respective resins can be reactivated by running weak solutions of hydrochloric acid and sodium hydroxide through them, reversing the equilibria represented above.

A catalyst with the potential to be used on the industrial scale should not only promote the formation of the desired product selectively. It also should be able to be confined physically within the reactor and at the end of the reaction it should be able to be recovered completely from the final mixture so that there is no contamination of the product by the catalyst. These requirements are not usually met easily in systems using homo-geneous (i.e. dissolved) catalysts, which only find restricted usage as a consequence, even if they are highly active and selective. These difficulties are overcome by the *immobilization* or *anchoring* of molecules of homogeneous catalyst to the surface of an insoluble solid. A comprehensive account of what has been achieved in this respect is well beyond the scope of this book, but a few examples should indicate the main principles and the enormous potential.

Immobilization is a term usually reserved for the attachment of enzymes to solid surfaces, while anchoring is applied most frequently when inorganic catalytic moieties are bound to the surface. The enzyme glucose oxidase is important for determining glucose levels in blood. The way in which the analysis is carried out requires the enzyme to be immobilized on a tough plastic membrane. Nylon 6,6 is suitable for this, after it has been subjected to the action of fairly strong hydrochloric acid under reflux conditions to hydrolyze some of the amide linkages thus

$$----C(=O)-N(H)---- \xrightarrow[\sim 10 \text{ mins}]{3M\ HCl(aq)} ----C(=O)-OH + H_2N----.$$

The enzyme (E) molecule is large and globular-shaped: it has amino groups at its surface but these are well removed from the small area (the active site) which is responsible for its catalytic action. The commonly used reagent used to create a chemical linkage between the amino groups is glutaraldehyde ($O=C(H)-(CH_2)_3-C(H)=O$), working as

$$E-NH_2 + OHC(CH_2)_3CHO + H_2N-nylon \rightarrow$$
$$E-N=CH(CH_2)_3CH=N-nylon + 2H_2O.$$

In fact, the immobilization of this enzyme was found to be optimized if hexamethylenediamine ($H_2N(CH_2)_6NH_2$) was coupled to the nylon amino group first to act as a 'spacer'. The enzyme was then coupled through this residue to the surface of the nylon membrane. The specific activity of the enzyme (i.e. the rate of the catalysed reaction per unit mass of the catalyst) was about a fifth of that achieved by the free (i.e. dissolved) enzyme under the same conditions. This is a very acceptable level of performance when the advantages of having the enzyme retained by the nylon membrane are considered. A 'biosensor' can then be constructed by wrapping this membrane loosely around an electrode which responds to one of the products of the enzyme-catalysed reaction (usually gluconic acid in the present case) but is insensitive to all other components of the medium (for example blood) being analysed.

Organotin compounds catalyse the reaction in which n-butanol adds to isocyanate to create a urethane linkage

$$n-C_4H_9OH + O=C=N-R \rightarrow n-C_4H_9-O-C(=O)-N(H)-R$$

where R represents the rest of the isocyanate molecule. The problem encountered in the homogeneous catalytic system is that 'deadlock' complexes, in which at least two of the organotin catalyst molecules are incorporated with several alcohol molecules, form and this loss of catalyst severely reduces the rate. Anchoring of the organtin molecules to a polymer surface is then an obvious means of keeping them apart and, in the resultant absence of deadlock complexes, reaction rates are achieved which are well above found in the corresponding homogeneous system. In one instance the support used was a copolymer of styrene with maleic anhydride

$$(O=C-C(H)=C(H)-C=O)$$
$$\underset{O}{\rule{3cm}{0.4pt}}$$

which had been synthesized by a radical addition mechanism. The organotin catalytic sites were produced by opening up the anhydride rings of the copolymer using $((C_4H_9)_3Sn)_2O$ at 333 K, to give a typical pair of noninteracting catalytic sites represented as $----CH(C(=O)-O-Sn(C_4H_9)_3)-CH(C(=O)-O-Sn(C_4H_9)_3)----$. A plot of the second order rate constant of the reaction against the amount of tin present in the anchored catalyst was linear, showing that there was no effective formation of deadlock complexes, even when the surface density of active sites was quite high. This is then a case in which anchoring of the catalyst is essential for obtaining acceptable activity, aside from the other advantages achieved. Hydroformylation of propylene yields *n*-butyraldehyde

$$CH_3-CH=CH_2 + CO + H_2 \rightarrow CH_3-CH_2-CH_2-C(H)=O.$$

This product is used mainly in the production of synthetic resins and plasticizers. Certain rhodium complexes offer high rates and selectivity as catalysts of this hydroformylation reaction, but these cannot be used in homogeneous systems because it is difficult to achieve complete recovery of the precious metal. This problem is resolved when the catalytic complexes, such as that represented as $RhCl(CO)(P(C_6H_5)_3)_2$ are anchored on an insoluble polymer, typically a crosslinked styrene–divinylbenzene copolymer which has been functionalized suitably, for instance with diphenylphosphine disulphonic acid $(-CH_2-P(C_6H_4-SO_3H)_2)$ groups. The resultant anchored catalysts induce rapid hydroformylation of propylene dissolving into an aqueous medium working at 373 K under pressures of 5–50 times normal atmospheric pressure.

Another pertinent example involves a catalyst which works as a Ziegler–Natta catalyst for ethylene polymerization (Section 7.4). The creation of the support commenced with polyethylene, onto which chains of the residues of allyl alcohol $(CH_2=CH-CH_2OH)$ were grafted via the action of γ-radiation from a ^{60}Co source (Section 9.1). The resulting graft copolymer was reacted with isobutyl magnesium chloride $(i-C_4H_9MgCl)$ and then titanium tetrachloride $(TiCl_4)$, to replace hydroxylic hydrogen atoms of $-CH_2OH$ substituents on the grafted chains with moieties, each of which is a complex containing magnesium and titanium atoms, the latter believed to be in the oxidation state Ti(III). These polymer-supported catalysts for ethylene polymerization present arrays of sites, active for catalytic action, which may be regarded as being similar to a monolayer, i.e. a layer one molecule thick across the surface. Their activities as catalysts are comparable with those shown by the other heterogeneous Ziegler–Natta catalysts discussed in Section 7.4.

Finally attention may be turned to perhaps the only commercial polymeric catalyst in current usage, sulphonated polystyrene resin. This can serve as an effective catalyst for most of the reactions of organic molecules which have been catalysed by sulphuric acid previously. There are considerable difficulties in using strong mineral acids on an industrial scale, including considerations of safety of workers and the corrosion of equipment. All of these problems cease when a solid resin with acidic function is used in place of a liquid acid. The major process concerned is the synthesis of methyl tertiary butyl ether (MTBE), used as an antiknock agent in gasoline fuels, from methanol and isobutene. The reaction may be represented as

$$CH_3OH + CH_2=C(CH_3)_2 \xrightarrow{----SO_3^- \ H^+} CH_3-O-C(CH_3)_3 \ (MTBE).$$

Other polymers functionalized with sulphonic acid groups are coming into commercial usage in other major processes, such as the productions of isopropanol (propan-2-ol) and n-butanol from the hydrations of propylene and 1-butene respectively. Of relevance to processes mentioned earlier are the actions of these solid acids in the production of bisphenol A (Section 9.4) and methylmethacrylate.

10.5 Concluding remarks

Space does not permit an extensive review of the exciting prospects for polymers, but it is to be hoped that the limited contents of the four preceding sections will have whetted the reader's appetite for some amazing developments in the years ahead. It is not so long ago that the suggestion that all of a vehicle, including its combustion engine, could be made of polymer-based materials would have been greeted with incredulity if not mirth. Now this is almost a reality. In future even more of the objects which enhance our lifestyles are likely to be made using these astonishingly versatile polymeric materials.

Finally it is hoped that this book will have succeeded in its aim of describing how polymers, tailored to achieve particular bulk properties, can be synthesized.

Appendix

Systematic names of common monomers

Common name	Systematic name
acetylene	ethyne
acrylonitrile	propenitrile
adipic acid	1,6-hexanedioic acid
bisphenol A	2,2-bis-4-hydroxyphenylpropane
butadiene	1,3-butadiene
n-butyl vinyl ether	1-(ethenyloxy)butane
caprolactam	2-oxohexamethyleneimine
chloroprene	2-chloro-1,3-butadiene
divinylbenzene	1,4-diethenylbenzene
ethylene	ethene
ethylene glycol	1,2-ethanediol
ethylene oxide	1,2-epoxyethane
formaldehyde	methanal
hexamethylene diamine	1,6-hexanediamine
isoprene	2-methyl-1,3-butadiene
maleic anhydride	butenedioic anhydride
melamine	2,4,6-triamino-s-triazine
methylmethacrylate	2-methylpropenoic acid methyl ester
phenol	hydroxybenzene
propylene	propene
propylene oxide	1,2-epoxypropane
pyrrole	pyrrole
styrene	ethenylbenzene
terephthalic acid	benzene-1,4-dicarboxylic acid
tetrafluoroethylene	tetrafluoroethene
urea	urea
vinyl acetate	3-butenoic acid ethenyl ester
vinyl alcohol	hydroxyethene
vinyl chloride	chloroethene
vinylidene fluoride	1,1'-difluoroethene

Further reading

The references below are grouped roughly on the basis of the chapter for which they have most significance apparently: but in some cases there will be relevance to topics involved in more than one chapter, so that the divisions should not be taken too rigorously.

General and Chapter 1

Alger, M. S. M. (1989) *Polymer science dictionary*, Elsevier, Barking.

Billmeyer, F. W. (1984) *Textbook of polymer science*, (3rd ed). Wiley, New York.

Carraher, C. A., Jr and Seymour, R. B. (1986) Physical aspects of polymer structure — a dictionary of terms, *Journal of Chemical Education*, **63**, 418-19.

Carraher, C. A., Jr and Seymour R. B. (1988) Polymer structure — organic aspects (definitions), *Journal of Chemical Education*, **65**, 314-19.

Cowie, J. M. G. (1991) *Polymers: chemistry and physics of modern materials*, (2nd edn). Blackie, Bishopbriggs.

Elias, H-G. (1987) *Megamolecules*, Springer, New York.

Kauffman, G. B. (1988) Teflon — 50 slippery years. *Education in Chemistry*, **25**, 173-5.

MacLachlan, A., (1990) Polymers: an industrial perspective, *Chemtech*, **20**, 590-3.

Seymour, R. B. (1988) Polymers are everywhere. *Journal of Chemical Education*, **65**, 327-34.

Steven, M. P. (1990) *Polymer chemistry*, (2nd edn). Oxford University Press, New York.

Chapter 2

Booth, C. and Price, C. (1989) *Comprehensive polymer science*, Pergamon Press, Oxford.

Hamielec, A. E. (1984) *Steric exclusion liquid chromatography of polymers*. Marcel Dekker, New York.

Matthias, L. J. (1983) Evaluation of a viscosity-molecular weight relationship, and the laboratory for introductory polymer courses. *Journal of Chemical Education*, **60**, 422-4 and 990-3.

Peebles, L. H. (1971) *Molecular weight distributions in polymers*. Interscience, New York.

Chapter 3

Allen, G. and Bevington, J. C. (ed.) (1989) *Comprehensive polymer science*, Vol. 2. Pergamon Press, Oxford.

Beck, K. R., Korsmeyer, R., and Kunz, R. J. (1984) An overview of the glass transition temperature of synthetic polymers. *Journal of Chemical Education*, **61**, 668-70.

Gordon, M. (1963) *High polymers*. Iliffe Books, London.

Hall, I. H. (ed.) (1984) *Structure of crystalline polymers*. Elsevier, Amsterdam.

Kaiser, T. (1989) Highly crosslinked polymers. *Progress in Polymer Science*, **14**, 373-450.

Mark, J. E., Eisenberg, A., Graessley, W. W., Mandelkern, L., and Koenig, J. L. (1984). *Physical properties of polymers*. American Chemical Society, Washington, DC.

Ward, I. M. (1983) *Mechanical properties of solid polymers*, (2nd edn). Wiley, Chichester, UK.

Chapter 4

Balta-Calleja, F. J. and Vonk, C. G. (1989) *X-ray scattering of synthetic polymers*. Elsevier, Amsterdam.

Klöpffer, W. (1984) *Introduction to polymer spectroscopy*. Springer, New York.

Mitchell, J. (ed.) (1987) *Applied polymer analysis and characterization: recent developments in techniques, instrumentation and problem solving*. Hanser, Munich.

Rabek, J. F. (1980) *Experimental methods in polymer chemistry*. Wiley-Interscience, New York.

White, J. R. (1989) *Polymer characterization*. Chapman & Hall, New York.

Chapter 5

Allen, G. and Bevington, J. C. (ed.) (1989) *Comprehensive polymer science*, Vol. 5. Pergamon Press, Oxford.

Faust, C. B. (1990) Silicone rubbers. *Education in chemistry*, **27**, 101-4.

Kamon, T. and Furukawa, H. (1986) Curing mechanisms and mechanical properties of cured epoxy resins. *Advances in Polymer Science*, **80**, 172-202.

Matthias, L. J., Vaidya, R. A., and Canterberry, J. B. (1984) Nylon 6 — a simple, safe synthesis of a tough commercial polymer. *Journal of Chemical Education*, **61**, 805-7.

Oertel, G. (ed.) (1985) *Polyurethane handbook*. Hanser, Munich.

Pizzi, A. (ed.) (1983) *Wood adhesives: chemistry and technology*, Marcel Dekker, New York.

Solomon, D. H. (ed.) (1973) *Step-growth polymerization*. Marcel Dekker, New York.

Chapter 6

Allen G. and Bevington, J. C. (ed.) (1989) *Comprehensive polymer science*, Vols 3 and 4. Pergamon Press, Oxford.

Bassett, D. R. and Hamielec, A. E. (ed.) (1981) *Emulsion polymers and emulsion polymerization*, ACS Symposium Series, **165**. American Chemical Society, Washington, DC.

Bevington, J. C. (1987) Initiation of polymerization: azo compounds and peroxides. *Makromolekulare Chemie, Macromolcular Symposia*, **10/11**, 89–107.

Bouton, T. C., Henderson, J. N., and Bevington, J. C. (ed.) (1979) *Polymerization reactors and processes*. ACS Symposium Series, **104**. American Chemical Society, Washington, DC.

Rempp, P. and Merrill, E. W. (1986) *Polymer synthesis*. Hüthig & Wepf, Basel.

Stannett, V. T. (1990) Radiation grafting — state-of-the-art. *Radiation Physics and Chemistry*, **35**, 82–7.

Stickler, M. (1987) Experimental techniques in free radical polymerization kinetics. *Makromolekulare Chemie, Macromolecular Symposia*, **10/11**, 17–69.

Tüdos, F. and Földes-Berezsnich, T. (1989) Free radical polymerization: inhibition and retardation. *Progress in Polymer Science*, **14**, 717–61.

Chapter 7

Allen, G. and Bevington, J. C. (ed.) (1989) *Comprehensive Polymer Science*, Vol. 4. Pergamon Press, Oxford.

Goodall, B. L. (1986) The history and current state of the art of propylene polymerization catalysts. *Journal of Chemical Education*, **63**, 191–95.

Grubbs, R. H. and Tumas, W. (1989) Polymer synthesis and organotransition metal chemistry. *Science*, **243**, 907–15.

Kaminsky, W. and Sinn, H. (ed.) (1988) *Transition metals and organometallics as catalysts for olefin polymerization*. Springer, Berlin.

Kennedy, J. P. and Marechal, E. (1982) *Carbocationic polymerization*. Wiley-Interscience, New York.

McGrath, J. E. (ed.) (1981) *Anionic polymerization: kinetics, mechanism and synthesis*. ACS Symposium Series, **166**, American Chemical Society, Washington, DC.

Morton, M. (1983) *Anionic polymerization: principles and practice*. Academic Press, New York.

Pino, P. and Moretti, G. (1987) The impact of the discovery of the polymerization of the α-olefines on the development of the stereospecific polymerization of vinyl monomers. *Polymer*, **28**, 683–92.

Rempp, P., Franta, E., and Herz, J. E. (1988) Macromolecular engineering by anionic methods. *Advances in Polymer Science*, **86**, 145–73.

Chapter 8

Agranoff, J. (ed.) (1984) *Modern plastics encyclopaedia*, 1984–85. McGraw-Hill, New York.

Decroly, P. (1987) Waste incineration: recovering the energy in the rubbish bin. *Plastics and Rubber International*, **12**, 32–5.

Klemchuk, P. P. (1990) Degradable plastics: a critical review. *Polymer Degradation and Stability*, **27**, 183–202.

Legge, N. R. (1989) Thermoplastic elastomers — three decades of progress. *Rubber Chemistry and Technology*, **62**, 529–47.

Lichstein, B. M. (1990) Recycling plastics and fibers. *Chemtech*, **20**, 360-4.

Mandelkern, L. (1964) *Crystallization of polymers.* McGraw–Hill, New York.

H. F. Mark, Bikales, N. M., Overberger, C. G., Menges, G., and Kroschwitz, J. I. (ed.) (1986) *Encyclopaedia of polymer science and engineering.* Wiley-Interscience, New York.

Morton-Jones, D. H. (1989) *Polymer processing.* Chapman & Hall, New York.

Chapter 9

Armand, M. (1990) Polymers with ionic conductivity *Advanced Materials*, **2**, 278-86.

Attard, G. and Imrie, C. (1991) Plastics of a new order. *New Scientist*, 11 May 1991, 38-43.

Ballauff, M. (1989) Stiff-chain polymers—structure, phase behavior and properties. *Angewandte Chemie* (International edition in English), **28**, 253-67.

Brostow, W. (1990) Properties of polymer liquid crystals: choosing molecular structures and blending. *Polymer*, **31**, 979-95.

Chanda, M. and Roy, S. K. (1987) *Plastics technology handbook.* Marcel Dekker, New York.

de Abajo, J. (1988) Modern trends in the synthesis and application of heat-resistant polymers. *Makromolekulare Chemie, Macromolecular Symposia*, **22**, 141-60.

Gaudiana, R. A., Minns, R. A., Sinta, R., Weeks, N., and Rogers, H. G. (1989) Amorphous rigid-rod polymers. *Progress in Polymer Science*, **14**, 47-89.

Grayson, M. (ed.) (1985) *Kirk–Othmer concise encyclopaedia of chemical technology.* Wiley, New York.

Richardson, T. L. (1989) *Industrial Plastics: Theory and Application.* Delmar, Albany.

Sommer, K. *et al.* (1991) Correlation between primary chemical structure and property phenomena in polycondensates. *Advanced Materials*, **3**, 590-9.

Vincent, C. A. (1989) Polymer electrolytes. *Chemistry in Britain*, **25**, 391-95.

Chapter 10

Attard, G. and Williams, G. (1986) Liquid crystalline side-chain polymers. *Chemistry in Britain*, **22**, 919-24.

Belbin, G. R. and Staniland, P. A. (1987) Advanced thermoplastics and their composites. *Philosophical Transactions of the Royal Society*, **A322**, 451-64.

Berlin, P. A., Levina, M. A., Tiger, R. P. and Entelis, S. G. (1991) Polymer-supported catalysts in nucleophilic addition of *n*-butanol to isocyanates. *Journal of Molecular Catalysis*, **64**, 15-22.

Birley, A. W., Haworth, B., and Batchelor, J. (1992) *Physics of plastics. Processing, properties and materials engineering.* Hanser, Munich.

Bochkin, A. M., Pomogailo, A. D, and Dyachkovskii, F. S. (1988) Polymer-supported titanium-magnesium catalysts for ethylene polymerization. *Reactive Polymers*, **9**, 99-107.

Chujo, Y., Nakamura, T., and Yamashita, Y. (1990) Synthesis of crown ether-terminated poly(methylmethacrylate) by radical chain transfer polymerization. *Journal of Polymer Science, Part A: Polymer Chemistry*, **28**, 59-65.

Da Silva, M. A., Gil, M. H., Redinha, J. S., Oliveira Brett, A. M., and Costa Pereira, J. L. (1991) Immobilization of glucose oxidase on nylon membranes and its application in a flow-through glucose reactor. *Journal of Polymer Science, Part A: Polymer Chemistry*, **29**, 275–9.

Dickstein, W. H. (1990) *Rigid rod star-block copolymers*. Technomic Publishing, Lancaster, USA.

McArdle, C. B. (ed.) (1989) *Side-chain liquid crystal polymers*. Blackie, Bishopbriggs.

Nair, C. P. R. and Clouet, G. (1990) Block copolymers via thermal polymeric iniferters. Synthesis of silicone–vinyl block copolymers. *Macromolecules*, **23**, 1361–9.

Ro, K. S. and Woo, S. I. (1991) Aqueous phase hydroformylation of propene catalysed over rhodium complexes immobilized on the poly(styrene–divinylbenzene) copolymer containing $-CH_2P(C_6H_4SO_3H)_2$ groups. *Applied Catalysis*, **69**, 169–75.

Sherrington, D. C. (1991) Polymer-supported systems: towards clean chemistry?. *Chemistry & Industry*, 7 January, 15–19.

Index